The Theory of Linear Prediction

The Theory of Linear Prediction

P. P. Vaidyanathan

ISBN: 978-3-031-01399-7   paperback
ISBN: 978-3-031-01399-7   paperback

ISBN: 978-3-031-02527-3   ebook
ISBN: 978-3-031-02527-3   ebook

DOI: 10.1007/978-3-031-02527-3

A Publication in the Springer series

SYNTHESIS LECTURES ON SIGNAL PROCESSING # 3

Lecture #3

Series Editor: José Moura, Carnegie Mellon University

Series ISSN

ISSN 1932-1236   print
ISSN 1932-1694   electronic

The Theory of Linear Prediction

P. P. Vaidyanathan
California Institute of Technology

SYNTHESIS LECTURES ON SIGNAL PROCESSING #3

To Usha, Vikram, and Sagar and my parents.

ABSTRACT

Linear prediction theory has had a profound impact in the field of digital signal processing. Although the theory dates back to the early 1940s, its influence can still be seen in applications today. The theory is based on very elegant mathematics and leads to many beautiful insights into statistical signal processing. Although prediction is only a part of the more general topics of linear estimation, filtering, and smoothing, this book focuses on linear prediction. This has enabled detailed discussion of a number of issues that are normally not found in texts. For example, the theory of vector linear prediction is explained in considerable detail and so is the theory of line spectral processes. This focus and its small size make the book different from many excellent texts which cover the topic, including a few that are actually dedicated to linear prediction. There are several examples and computer-based demonstrations of the theory. Applications are mentioned wherever appropriate, but the focus is not on the detailed development of these applications. The writing style is meant to be suitable for self-study as well as for classroom use at the senior and first-year graduate levels. The text is self-contained for readers with introductory exposure to signal processing, random processes, and the theory of matrices, and a historical perspective and detailed outline are given in the first chapter.

KEYWORDS

Linear prediction theory, vector linear prediction, linear estimation, filtering, smoothing, line spectral processes, Levinson's recursion, lattice structures, autoregressive models

Preface

Linear prediction theory has had a profound impact in the field of digital signal processing. Although the theory dates back to the early 1940s, its influence can still be seen in applications today. The theory is based on very elegant mathematics and leads to many beautiful insights into statistical signal processing. Although prediction is only a part of the more general topics of linear estimation, filtering, and smoothing, I have focused on linear prediction in this book. This has enabled me to discuss in detail a number of issues that are normally not found in texts. For example, the theory of vector linear prediction is explained in considerable detail and so is the theory of line spectral processes. This focus and its small size make the book different from many excellent texts that cover the topic, including a few that are actually dedicated to linear prediction. There are several examples and computer-based demonstrations of the theory. Applications are mentioned wherever appropriate, but the focus is not on the detailed development of these applications.

The writing style is meant to be suitable for self-study as well as for classroom use at the senior and first-year graduate levels. Indeed, the material here emerged from classroom lectures that I had given over the years at the California Institute of Technology. So, the text is self-contained for readers with introductory exposure to signal processing, random processes, and the theory of matrices. A historical perspective and a detailed outline are given in Chapter 1.

Acknowledgments

The pleasant academic environment provided by the California Institute of Technology and the generous support from the National Science Foundation and the Office of Naval Research have been crucial in developing some of the advanced materials covered in this book.

During my "young" days, I was deeply influenced by an authoritative tutorial on linear filtering by Prof. Tom Kailath (1974) and a wonderful tutorial on linear prediction by John Makhoul (1975). These two articles, among other excellent references, have taught me a lot and so has the book by Anderson and Moore (1979). My "love for linear prediction" was probably kindled by these three references. The small contribution I have made here would not have been possible were it not for these references and other excellent ones mentioned in the introduction in Chapter 1.

It is impossible to reduce to words my gratitude to Usha, who has shown infinite patience during my busy days with research and book projects. She has endured many evenings and weekends of my "disappearance" to work. Her sincere support and the enthusiasm and love from my uncomplaining sons Vikram and Sagar are much appreciated!

P. P. Vaidyanathan

California Institute of Technology

Contents

CHAPTER 1

Introduction

Digital signal processing has influenced modern technology in many wonderful ways. During the course of signal processing history, several elegant theories have evolved on a variety of important topics. These theoretical underpinnings, of which we can be very proud, are certainly at the heart of the crucial contributions that signal processing has made to the modern technological society we live in.

One of these is the theory of linear prediction. This theory, dating back to the 1940s, is fundamental to a number of signal processing applications. For example, it is at the center of many modern power spectrum estimation techniques. A variation of the theory arises in the identification of the direction of arrival of an electromagnetic wave, which is important in sensor networks, array processing, and radar. The theory has been successfully used for the representation, modeling, compression, and computer generation of speech waveforms. It has also given rise to the idea of line spectrum pairs, which are used in speech compression based on perceptual measures. More recently, vector versions of linear prediction theory have been applied for the problem of blind identification of noisy communication channels.

Another product of the theory is a class of filtering structures called *lattice structures*. These have been found to be important in speech compression. The same structures find use in the design of robust adaptive digital filter structures. The infinite impulse response (IIR) version of the linear prediction lattice is identical to the well-known all-pass lattice structure that arises in digital filter theory. The lattice has been of interest because of its stability and robustness properties despite quantization.

In this book, we give a detailed presentation of the theory of linear prediction and place in evidence some of the applications mentioned above. Before discussing the scope and outline, it is important to have a brief glimpse of the history of linear prediction.

1.1   HISTORY OF LINEAR PREDICTION

Historically, linear prediction theory can be traced back to the 1941 work of Kolmogorov, who considered the problem of extrapolation of discrete time random processes. Other early pioneers are Levinson (1947), Wiener (1949), and Weiner and Masani (1958), who showed how to extend the ideas for the case of multivariate processes. One of Levinson's contributions, , which for some reason he regarded as "mathematically trivial," is still in wide use today (Levinson's recursion). For a very detailed scholarly review of the history of statistical filtering, we have to refer the reader to the classic article by Kailath (1974), wherein the history of linear estimation is traced back to its very roots.

An influential early tutorial is the article by John Makhoul (1975) , which reviews the mathematics of linear prediction, Levinson's recursion, and so forth and makes the connection to spectrum estimation and the representation of speech signals. The connection to power spectrum estimation is studied in great detail in a number of early articles [Kay and Marple, 1981; Robinson, 1982; Marple, 1987; Kay, 1988]. Connections to maximum entropy and spectrum estimation techniques can be found in Kay and Marple (1981), Papoulis (1981), and Robinson (1982). Articles that show the connection to direction of arrival and array processing include Schmidt (1979), Kumaresan (1983), and Paulraj et al. (1986).

Pioneering work that explored the application in speech coding includes the work of Atal and Schroeder (1970) and that of Itakura and Saito (1970). Other excellent references for this are Makhoul (1975), Rabiner and Schafer (1978), Jayant and Noll (1984), Deller et al. (1993), and Schroeder (1999). Applications of multivariable models can be found even in the early image processing literature (e.g., see Chellappa and Kashyap, 1985).

The connection to lattice structures was studied by Gray and Markel (1973) and Makhoul (1977), and their applications in adaptive filtering and channel equalization were studied by a number of authors (e.g., Satorius and Alexander, 1979). This application can be found in a number of books (e.g., Haykin, 2002; Sayed, 2003). All-pass lattice structures that arise in digital filter theory are explained in detail in standard signal processing texts (Vaidyanathan, 1993; Proakis and Manolakis, 1996; Oppenheim and Schafer, 1999; Mitra, 2001; Antoniou, 2006).

Although the history of linear prediction can be traced back to the 1940s, it still finds new applications. This is the beauty of any solid mathematical theory. An example is the application of the vector version of linear prediction theory in blind identification of noisy finite impulse response (FIR) channels (Gorokhov and Loubaton, 1999; Lopez-Valcarce and Dasgupta, 2001).

1.2   SCOPE AND OUTLINE

Many signal processing books include a good discussion of linear prediciton theory, for example, Markel and Gray (1976), Rabiner and Schafer (1978), Therrien (1992), Deller et al. (1993), Anderson and Moore (1979), Kailath et al. (2000), Haykin (2002), and Sayed (2003). The book by Strobach (1990) is dedicated to an extensive discussion of linear prediction, and so is the early book by Markel and Gray (1976), which also discusses application to speech. The book by Therrien (1992) not only has a nice chapter on linear prediction, it also explains the applications in spectrum estimation extensively. Another excellent book is the one by Kailath et al. (2000), which focuses on the larger topic of linear estimation. Connections to array processing are also covered by Therrien (1992) and Van Trees (2002).

This short book focuses on the theory of linear prediction. The tight focus allows us to include under one cover details that are not normally found in other books. The style and emphasis here are different from the above references. Unlike most of the other references, we have included a thorough treatment of the vector case (multiple-input/multiple-output linear predictive coding, or MIMO LPC) in a separate chapter. Another novelty is the inclusion of a detailed chapter on the theory of line spectral processes. There are several examples and computer-based demonstrations throughout the book to enhance the theoretical ideas. Applications are briefly mentioned so that the reader can see the connections. The reader interested in these applications should peruse some of the references mentioned above and references in the individual chapters.

Chapter 2 introduces the optimal linear prediction problem and develops basic equations for optimality, called the *normal equations*. A number of properties of the solution are also studied. Chapter 3 introduces Levinson's recursion, which is a fast procedure to solve the normal equations. This recursion places in evidence further properties of the solution. In Chapter 4, we develop lattice structures for linear prediction. Chapter 5 is dedicated to the topic of autoregressive modeling, which has applications in signal representation and compression.

In Chapter 6, we develop the idea of flatness of a power spectrum and relate it to predictability of a process. Line spectral processes are discussed in detail in Chapter 7. One application is in the identification of sinusoids in noise, which is similar to the problem of identifying the direction of arrival of an electromagnetic wave. The chapter also discusses the theory of line spectrum pairs, which are used in speech compression.

In Chapter 8, a detailed discussion of linear prediction for the MIMO case (case of vector processes) is presented. The MIMO lattice and its connection to paraunitary matrices is also explored in this chapter. The appendices include some review material on linear estimation as well as a few details pertaining to some of the proofs in the main text.

A short list of homework problems covering most of the chapters is included at the end, before the bibliography section.

1.2.1 Notations

Boldface letters such as $\mathbf{A}$ and $\mathbf{v}$ indicate matrices and vectors. Superscript T,*, and †as in $\mathbf{A}^{\mathrm{T}}$, $\mathbf{A}^*$, and $\mathbf{A}^{\dagger}$ denote, respectively, the transpose, conjugate, and transpose–conjugate of a matrix. The tilde notation on a function of z is defined as follows:

$$\widetilde{\mathbf{H}}(z) = \mathbf{H}^{\dagger}(1/z^*)$$

Thus,

$$\mathbf{H}(z) = \sum_n \mathbf{h}(n)z^{-n} \Rightarrow \widetilde{\mathbf{H}}(z) = \sum_n \mathbf{h}^\dagger(n)z^n,$$

so that the tilde notation effectively replaces all coefficients with the transpose conjugates and replaces z with $1/z$. For example,

$$H(z) = h(0) + h(1)z^{-1} \Rightarrow \widetilde{H}(z) = h^*(0) + h^*(1)z$$

and

$$H(z) = \frac{a_0 + a_1 z^{-1}}{1 + b_1 z^{-1}} \Rightarrow \widetilde{H}(z) = \frac{a_0^* + a_1^* z}{1 + b_1^* z}.$$

Note that the tilde notation reduces to transpose conjugation on the unit circle:

$$\widetilde{\mathbf{H}}(e^{j\omega}) = \mathbf{H}^\dagger(e^{j\omega})$$

As mentioned in the preface, the text is self-contained for readers with introductory exposure to signal processing, random processes, and matrices. The determinant of a square matrix $\mathbf{A}$ is denoted as $\det \mathbf{A}$ and the trace as $\mathrm{Tr}\,\mathbf{A}$. Given two Hermitian matrices $\mathbf{A}$ and $\mathbf{B}$, the notation $\mathbf{A} \geq \mathbf{B}$ means that $\mathbf{A} - \mathbf{B}$ is positive semidefinite, and $\mathbf{A} > \mathbf{B}$ means that $\mathbf{A} - \mathbf{B}$ is positive definite.

•   •   •   •   •

CHAPTER 2

The Optimal Linear Prediction Problem

2.1 INTRODUCTION

In this chapter, we introduced the optimal linear prediction problem. We develop the equations for optimality and discuss some properties of the solution.

2.2 PREDICTION ERROR AND PREDICTION POLYNOMIAL

Let $x(n)$ be a wide sense stationary (WSS) random process (Papoulis, 1965), possibly complex. Suppose we wish to predict the value of the sample $x(n)$ using a linear combination of N most recent past samples. The estimate has the form

$$\widehat{x}_N^f(n) = -\sum_{i=1}^{N} a_{N,i}^* \, x(n-i). \qquad (2.1)$$

The integer N is called the prediction order. Notice the use of two subscripts for a, the first one being the prediction order. The superscript f on the left is a reminder that we are discussing the "forward" predictor, in contrast to the "backward predictor" to be introduced later. The estimation error is

$$e_N^f(n) = x(n) - \widehat{x}_N^f(n), \qquad (2.2)$$

that is,

$$e_N^f(n) = x(n) + \sum_{i=1}^{N} a_{N,i}^* \, x(n-i). \qquad (2.3)$$

We denote the *mean squared error* as $\mathcal{E}_N^f$:

$$\mathcal{E}_N^f \triangleq E[|e_N^f(n)|^2]. \qquad (2.4)$$

In view of WSS property, this is independent of time. The optimum predictor (i.e., the optimum set of coefficients $a_{N,i}^*$) is the one that minimizes this mean squared value. From Eq. (2.3), we see

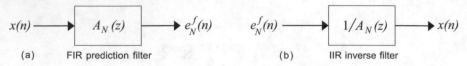

FIGURE 2.1: (a) The FIR prediction filter and (b) its inverse.

that the prediction error $e_N^f(n)$ can be regarded as the output of an FIR filter $A_N(z)$ in response to the WSS input $x(n)$. See Fig. 2.1. The FIR filter transfer function is given by

$$A_N(z) = 1 + \sum_{i=1}^{N} a_{N,i}^* \, z^{-i}. \qquad (2.5)$$

The IIR filter $1/A_N(z)$ can therefore be used to reconstruct $x(n)$ from the error signal $e_N^f(n)$ (Fig. 2.1(b)). The conjugate sign on $a_{N,i}$ in the definition (2.5) is for future convenience. Because $e_N^f(n)$ is the output of a filter in response to a WSS input, we see that $e_N^f(n)$ is itself a WSS random process. $A_N(z)$ is called the *prediction polynomial*, although its output is only the prediction error.

Thus, linear prediction essentially converts the signal $x(n)$ into the set of N numbers $\{a_{N,i}\}$ and the error signal $e_N^f(n)$. We will see later that the error $e_N^f(n)$ has a relatively flat power spectrum compared with $x(n)$. For large N, the error is nearly white, and the spectral information of $x(n)$ is mostly contained in the coefficient $\{a_{N,i}\}$. This fact is exploited in data compression applications. The technique of *linear predictive coding* (LPC) is the process of converting segments of a real time signal into the small set of numbers $\{a_{N,i}\}$ for storage and transmission (Section 5.6).

2.3 THE NORMAL EQUATIONS

From Appendix A.1, we know that the optimal value of $a_{N,i}$ should be such that the error $e_N^f(n)$ is *orthogonal* to $x(n - i)$, that is,

$$E[e_N^f(n)x^*(n - i)] = 0, \quad 1 \le i \le N. \qquad (2.6)$$

This condition gives rise to N equations similar to Eq. (A.13). The elements of the matrices **R** and **r**, defined in Eq. (A.13), now have a special form. Thus,

$$[\mathbf{R}]_{im} = E[x(n - 1 - i)x^*(n - 1 - m)], \quad 0 \le i, m \le N - 1.$$

Define $R(k)$ to be the autocorrelation sequence of the WSS process $x(n)$, that is,[1]

$$R(k) = E[x(n)x^*(n-k)]. \tag{2.7}$$

Using the fact that $R(k) = R^*(-k)$, we can then simplify Eq. (A.13) to obtain

$$\underbrace{\begin{bmatrix} R(0) & R(1) & \dots R(N-1) \\ R^*(1) & R(0) & \dots R(N-2) \\ \vdots & \vdots & \ddots & \vdots \\ R^*(N-1) & R^*(N-2) & \dots & R(0) \end{bmatrix}}_{\mathbf{R}_N} \begin{bmatrix} a_{N,1} \\ a_{N,2} \\ \vdots \\ a_{N,N} \end{bmatrix} = -\underbrace{\begin{bmatrix} R^*(1) \\ R^*(2) \\ \vdots \\ R^*(N) \end{bmatrix}}_{-\mathbf{r}} \tag{2.8}$$

For example, with $N = 3$, we get

$$\underbrace{\begin{bmatrix} R(0) & R(1) & R(2) \\ R^*(1) & R(0) & R(1) \\ R^*(2) & R^*(1) & R(0) \end{bmatrix}}_{\mathbf{R}_3} \begin{bmatrix} a_{3,1} \\ a_{3,2} \\ a_{3,3} \end{bmatrix} = -\begin{bmatrix} R^*(1) \\ R^*(2) \\ R^*(3) \end{bmatrix}. \tag{2.9}$$

These equations have been known variously in the literature as *normal equations*, *Yule–Walker equations*, and *Wiener–Hopf equations*. We shall refer to them as normal equations. We can find a unique set of optimal predictor coefficients $a_{N,i}$, as long as the $N \times N$ matrix $\mathbf{R}_N$ is nonsingular. Singularity of this matrix will be analyzed in Section 2.4.2. Note that the matrix $\mathbf{R}_N$ is *Toeplitz*, that is, all the elements on any line parallel to the main diagonal are identical. We will elaborate more on this later.

Minimum-phase property. The optimal predictor polynomial $A_N(z)$ in Eq. (2.5), which is obtained from the solution to the normal equations, has a very interesting property. Namely, all its zeros, z_k satisfies $|z_k| \leq 1$. That is, $A_N(z)$ is a mimimum-phase polynomial. In fact, the zeros are strictly inside the unit circle $|z_k| < 1$, unless $x(n)$ is a very restricted process called a line spectral process (Section 2.4.2). The minimum-phase property guarantees that the IIR filter $1/A_N(z)$ is stable. The proof of the minimum-phase property follows automatically as a corollary of the so-called Levinson's recursion, which will be presented in Section 3.2. A more direct proof is given in Appendix C. □

---

[1]A more appropriate notation would be $R_{xx}(k)$, but we have omitted the subscript for simplicity.

2.3.1   Expression for the Minimized Mean Square Error

We can use Eq. (A.17) to arrive at the following expression for the minimized mean square error:

$$\mathcal{E}_N^f = R(0) + \sum_{i=1}^{N} a_{N,i}^* \, R^*(i).$$

Because $\mathcal{E}_N^f$ is real, we can conjugate this to obtain

$$\mathcal{E}_N^f = R(0) + \sum_{i=1}^{N} a_{N,i} R(i). \tag{2.10}$$

Next, because $e_N^f(n)$ is orthogonal to the past N samples $x(n-k)$, it is also orthogonal to the estimate $\widehat{x}_N^f(n)$. Thus, from $x(n) = \widehat{x}_N^f(n) + e_N^f(n)$, we obtain

$$E[\, |x(n)|^2\,] = E[\, |\widehat{x}(n)|^2\,] + \underbrace{E[\, |e_N^f(n)|^2\,]}_{\mathcal{E}_N^f}. \tag{2.11}$$

Thus the mean square value of $x(n)$ is the sum of mean square values of the estimate and the estimation error.

Example 2.1: Second-Order Optimal Predictor. Consider a real WSS process with auto-correlation sequence

$$R(k) = (24/5) \times 2^{-|k|} - (27/10) \times 3^{-|k|}. \tag{2.12}$$

The values of the first few coefficients of $R(k)$ are

$$R(0) = 2.1, \; R(1) = 1.5, \; R(2) = 0.9, \; \ldots \tag{2.13}$$

The first-order predictor produces the estimate

$$\widehat{x}_1^f(n) = -a_{1,1}x(n-1), \tag{2.14}$$

and the optimal value of $a_{1,1}$ is obtained from $R(0)a_{1,1} = -R(1)$, that is,

$$a_{1,1} = -\frac{R(1)}{R(0)} = -\frac{5}{7}$$

The optimal predictor polynomial is

$$A_1(z) = 1 + a_{1,1}z^{-1} = 1 - (5/7)z^{-1} \tag{2.15}$$

The minimized mean square error is, from Eq. (2.10),

$$\mathcal{E}_1^f = R(0) + a_{1,1}R(1) = 36/35. \tag{2.16}$$

The second-order optimal predictor coefficients are obtained by solving the normal equations Eq. (2.8) with $N = 2$, that is,

$$\begin{bmatrix} 2.1 & 1.5 \\ 1.5 & 2.1 \end{bmatrix} \begin{bmatrix} a_{2,1} \\ a_{2,2} \end{bmatrix} = -\begin{bmatrix} 1.5 \\ 0.9 \end{bmatrix} \tag{2.17}$$

The result is $a_{2,1} = -5/6$ and $a_{2,2} = 1/6$. Thus, the optimal predictor polynomial is

$$A_2(z) = 1 - (5/6)z^{-1} + (1/6)z^{-2}. \tag{2.18}$$

The minimized mean square error, computed from Eq. (2.10), is

$$\mathcal{E}_2^f = R(0) + a_{2,1}R(1) + a_{2,2}R(2) = 1.0. \tag{2.19}$$

2.3.2 The Augmented Normal Equation

We can augment the error information in Eq. (2.10) to the normal equations (2.8) simply by moving the right hand side in Eq. (2.8) to the left and adding an extra row at the top of the matrix $\mathbf{R}_N$. The result is

$$\underbrace{\begin{bmatrix} R(0) & R(1) & \ldots & R(N) \\ R^*(1) & R(0) & \ldots & R(N-1) \\ \vdots & \vdots & \ddots & \vdots \\ R^*(N) & R^*(N-1) & \ldots & R(0) \end{bmatrix}}_{\mathbf{R}_{N+1}} \underbrace{\begin{bmatrix} 1 \\ a_{N,1} \\ \vdots \\ a_{N,N} \end{bmatrix}}_{\mathbf{a}_N} = \begin{bmatrix} \mathcal{E}_N^f \\ 0 \\ \vdots \\ 0 \end{bmatrix} \tag{2.20}$$

This is called the *augmented normal equation* for the Nth-order optimal predictor and will be used in many of the following sections. We conclude this section with a couple of remarks about normal equations.

1. *Is any Toeplitz matrix an autocorrelation?* Premultiplying both sides of the augmented equation by vector $\mathbf{a}_N^\dagger$, we obtain

$$\mathbf{a}_N^\dagger \mathbf{R}_{N+1} \mathbf{a}_N = \mathcal{E}_N^f. \tag{2.21}$$

Because $\mathbf{R}_{N+1}$ is positive semidefinite, this gives a second verification of the (obvious) fact that $\mathcal{E}_N^f \geq 0$. This observation, however, has independent importance. It can be used to prove that *any positive definite Toeplitz matrix* is an autocorrelation of some WSS process (see Problem 14).

2. *From predictor coefficients to autocorrelations.* Given the set of autocorrelation coefficients $R(0), R(1), \ldots, R(N)$, we can uniquely identify the N predictor coefficients $a_{N,i}$ and the mean square error $\mathcal{E}_N^f$ by solving the normal equations (assuming nonsingularity) and then using Eq. (2.10). Conversely, suppose we are given the solution $a_{N,i}$ and the error $\mathcal{E}_N^f$. Then, we can work backward and uniquely identify the autocorrelation coefficients $R(k)$, $0 \le k \le N$. This result, perhaps not obvious, is justified as part of the proof of Theorem 5.2 later.

2.4   PROPERTIES OF THE AUTOCORRELATION MATRIX

The $N \times N$ autocorrelation matrix $\mathbf{R}_N$ of a WSS process can be written as

$$\mathbf{R}_N = E[\mathbf{x}(n)\mathbf{x}^\dagger(n)] \tag{2.22}$$

where

$$\mathbf{x}(n) = \begin{bmatrix} x(n) & x(n-1) & \ldots & x(n-N+1) \end{bmatrix}^{\mathrm{T}}. \tag{2.23}$$

For example, the 3×3 matrix $\mathbf{R}_3$ in Eq. (2.9) is

$$\mathbf{R}_3 = E \begin{bmatrix} x(n) \\ x(n-1) \\ x(n-2) \end{bmatrix} \begin{bmatrix} x^*(n) & x^*(n-1) & x^*(n-2) \end{bmatrix}.$$

This follows from the definition $R(k) = E[x(n)x^*(n-k)]$ and from the property $R(k) = R^*(-k)$. We first observe some of the simple properties of $\mathbf{R}_N$:

1. *Positive definiteness.* Because $\mathbf{x}(n)\mathbf{x}^\dagger(n)$ is Hermitian and positive semidefinite, $\mathbf{R}_N$ is also Hermitian and positive semidefinite. In fact, it is positive *definite* as long as it is nonsingular. Singularity is discussed in Section 2.4.2.

2. *Diagonal elements.* The quantity $R(0)$ appears on all the diagonal elements and is the mean square value of the random process, that is, $R(0) = E[|x(n)|^2]$.

3. *Toeplitz property.* We observed earlier that $\mathbf{R}_N$ has the Toeplitz property, that is, the (k, m) element of $\mathbf{R}_N$ depends only on the difference $m - k$. To prove the Toeplitz property formally, simply observe that

$$
\begin{aligned}
[\mathbf{R}_N]_{km} &= E[x(n - k)x^*(n - m)] \\
&= E[x(n)x^*(n - m + k)] \\
&= R(m - k).
\end{aligned}
$$

The Toeplitz property is a consequence of the WSS property of $x(n)$. In Section 3.2, we will derive a fast procedure called the *Levinson's recursion* to solve the normal equations. This recursion is made possible because of the Toeplitz property of $\mathbf{R}_N$.

4. *A filtering interpretation of eigenvalues.* The eigenvalues of the Toeplitz matrix $\mathbf{R}_N$ can be given a nice interpretation. Consider an FIR filter

$$
V(z) = v_0^* + v_1^* z^{-1} + \ldots + v_{N-1}^* \, z^{-(N-1)} \tag{2.24}
$$

with input $x(n)$. Its output can be expressed as

$$
y(n) = v_0^* x(n) + v_1^* x(n - 1) + \ldots + v_{N-1}^* \, x(n - N + 1) = \mathbf{v}^\dagger \mathbf{x}(n),
$$

where

$$
\mathbf{v}^\dagger = \begin{bmatrix} v_0^* & v_1^* & \cdots & v_{N-1}^* \end{bmatrix}
$$

The mean square value of $y(n)$ is $E[|\mathbf{v}^\dagger \mathbf{x}(n)|^2] = \mathbf{v}^\dagger E[\mathbf{x}(n)\mathbf{x}^\dagger(n)]\mathbf{v} = \mathbf{v}^\dagger \mathbf{R}_N \mathbf{v}$. Thus,

$$
E[|y(n)|^2] = \mathbf{v}^\dagger \mathbf{R}_N \mathbf{v}. \tag{2.25}
$$

That is, given the Toeplitz autocorrelation matrix $\mathbf{R}_N$, the quadratic form $\mathbf{v}^\dagger \mathbf{R}_N \mathbf{v}$ is the mean square value of the output of the FIR filter $V(z)$, with input signal $x(n)$. In particular, let $\mathbf{v}$ be an unit-norm eigenvector of $\mathbf{R}_N$ with eigenvalue λ, that is,

$$
\mathbf{R}_N \mathbf{v} = \lambda \mathbf{v}, \quad \mathbf{v}^\dagger \mathbf{v} = 1.
$$

Then,

$$
\mathbf{v}^\dagger \mathbf{R}_N \mathbf{v} = \lambda. \tag{2.26}
$$

Thus, any eigenvalue of $\mathbf{R}_N$ can be regarded as the mean square value of the output $y(n)$ for appropriate choice of the unit-energy filter $V(z)$ driven by $x(n)$.

In the next few subsections, we study some of the deeper properties of the autocorrelation matrix. These will be found to be useful for future discussions.

2.4.1   Relation Between Eigenvalues and the Power Spectrum

Suppose λ_i, $0 \leq i \leq N-1$ are the eigenvalues of $\mathbf{R}_N$. Because the matrix is positive semidefinite, we know that these are real and nonnegative. Now, let $S_{xx}(e^{j\omega})$ be the power spectrum of the process $x(n)$, that is,

$$S_{xx}(e^{j\omega}) = \sum_{k=-\infty}^{\infty} R(k)e^{-j\omega k}. \qquad (2.27)$$

We know $S_{xx}(e^{j\omega}) \geq 0$. Now, let S_{min} and S_{max} denote the extreme values of the power spectrum (Fig. 2.2). We will show that the eigenvalues λ_i are bounded as follows:

$$S_{min} \leq \lambda_i \leq S_{max}. \qquad (2.28)$$

For example, if $x(n)$ is zero-mean white, then $S_{xx}(e^{j\omega})$ is constant, and all the eigenvalues are equal (consistent with the fact that $\mathbf{R}_N$ is diagonal with all diagonal elements equal to $R(0)$).

Proof of Eq. (2.28). Consider again a filter $V(z)$ as in (2.24) with input $x(n)$ and denote the output as $y(n)$. Because the mean square value is the integral of the power spectrum, we have

$$\begin{aligned}
E[|y(n)|^2] &= \frac{1}{2\pi} \int_0^{2\pi} S_{yy}(e^{j\omega})d\omega \\
&= \frac{1}{2\pi} \int_0^{2\pi} S_{xx}(e^{j\omega})|V(e^{j\omega})|^2 d\omega \\
&\leq S_{max} \int_0^{2\pi} |V(e^{j\omega})|^2 \frac{d\omega}{2\pi} = S_{max}\mathbf{v}^\dagger\mathbf{v}
\end{aligned}$$

where the last equality follows from Parseval's theorem. Similarly, $E[|y(n)|^2] \geq S_{min}\mathbf{v}^\dagger\mathbf{v}$. With $\mathbf{v}$ constrained to have unit norm ($\mathbf{v}^\dagger\mathbf{v} = 1$), we then have

$$S_{min} \leq E[|y(n)|^2] \leq S_{max} \qquad (2.29)$$

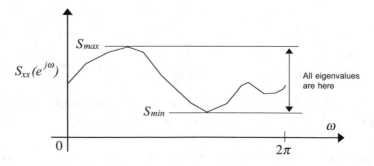

FIGURE 2.2: The relation between power spectrum and eigenvalues of the autocorrelation matrix.

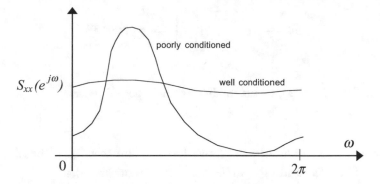

FIGURE 2.3: Examples of power spectra that are well and poorly conditioned.

But we have already shown that any eigenvalue of $\mathbf{R}_N$ can be regarded as the mean square value $E[|y(n)|^2$ for appropriate choice of the unit-energy filter $V(z)$ (see remark after Eq. (2.26)). Because (2.29) holds for every possible output $y(n)$ with unit-energy $V(z)$, Eq. (2.28), therefore, follows. □

With $\lambda_{\min}$ and $\lambda_{\max}$ denoting the extreme eigenvalues of $\mathbf{R}_N$, the ratio

$$\mathcal{N} = \frac{\lambda_{\max}}{\lambda_{\min}} \geq 1 \qquad (2.30)$$

is called the *condition number* of the matrix. It is well-known (Golub and Van Loan, 1989) that if this ratio is large, then numerical errors tend to have a more severe effect during matrix inversion (or when trying to solve Eq. (2.8)). It is also known that the condition number cannot decrease as the size N increases (Problem 12). If the condition number is close to unity, we say that the system of equations is well-conditioned. By comparison with Eq. (2.28), we see that

$$1 \leq \mathcal{N} = \frac{\lambda_{\max}}{\lambda_{\min}} \leq \frac{S_{\max}}{S_{\min}} \qquad (2.31)$$

Thus, if a random process has a nearly flat power spectrum (i.e., $S_{\max}/S_{\min} \approx 1$), it can be considered to be a well-conditioned process. If the power spectrum has a wide range, it is possible that the process is poorly conditioned (i.e., $\mathcal{N}$ could be very large). See Fig. 2.3 for demonstration. Also see Problem 11.

2.4.2 Singularity of the Autocorrelation Matrix

If the autocorrelation matrix is singular, then the corresponding random variables are linearly

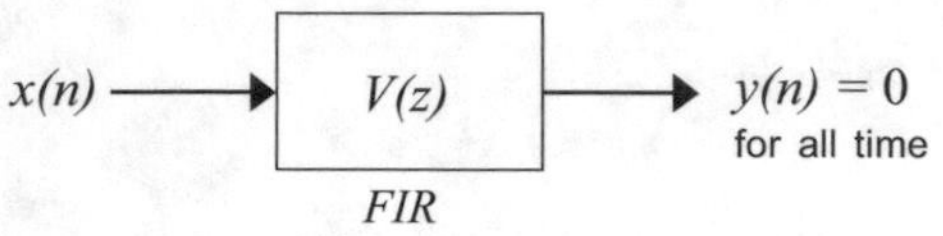

FIGURE 2.4: The FIR filter $V(z)$, which annihilates a line spectral process.

dependent (Section A.4). To be more quantitative, let us assume that $\mathbf{R}_{L+1}$ [the $(L+1) \times (L+1)$ matrix] is singular. Then, proceeding as in Section A.4, we conclude that

$$v_0^* \, x(n) + v_1^* \, x(n-1) + \ldots + v_L^* \, x(n-L) = 0, \tag{2.32}$$

where not all v_i's are zero. This means, in particular, that if we measure a set of L successive samples of $x(n)$, then all the future samples can be computed recursively, with no error. That is, the process is *fully predictable*.

Eq. (2.32) implies that if we pass the WSS random process $x(n)$ through an FIR filter (see Fig. 2.4),

$$V(z) = v_0^* + v_1^* \, z^{-1} + \ldots + v_L^* \, z^{-L}, \tag{2.33}$$

then the output $y(n)$ is zero for all time! Its power spectrum $S_{yy}(e^{j\omega})$, therefore, is zero for all ω. Thus

$$S_{yy}(e^{j\omega}) = S_{xx}(e^{j\omega})|V(e^{j\omega})|^2 \equiv 0. \tag{2.34}$$

Because $V(z)$ has at most L zeros on the unit circle (i.e., $|V(e^{j\omega})|^2$ has, at most, L distinct zeros in $0 \le \omega < 2\pi$), we conclude that $S_{xx}(e^{j\omega})$ can be nonzero only at these points. That is, it has the form

$$S_{xx}(e^{j\omega}) = 2\pi \sum_{i=1}^{L} c_i \delta_a(\omega - \omega_i), \quad 0 \le \omega < 2\pi, \tag{2.35}$$

which is a linear combination of Dirac delta functions. This is demonstrated in Fig. 2.5. The autocorrelation of the process $x(n)$, which is the inverse Fourier transform of $S_{xx}(e^{j\omega})$, then takes the form

$$R(k) = \sum_{i=1}^{L} c_i e^{j\omega_i k}. \tag{2.36}$$

A WSS process characterized by the power spectrum Eq. (2.35) (equivalently the autocorrelation (2.36)) is said to be a *line spectral process*, and the frequencies ω_i are called the *line frequencies*.

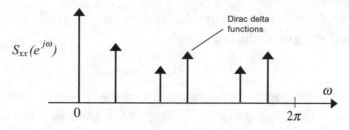

FIGURE 2.5: Power spectrum of a line spectral process.

Because the process is fully predictable, it is the exact "opposite" of a *white* process, which has no correlation between any pair of samples. In terms of frequency domain, for a white process, the power spectrum is constant, whereas for a fully predictable process, the power spectrum can only have impulses—it cannot have any smooth component.

In Chapter 7, we present a more complete study of these and address the problem of identifying the parameters ω_i and c_i when such a process is buried in noise.

2.4.3 Determinant of the Autocorrelation Matrix

We now show that the minimized mean square error $\mathcal{E}_N^f$ can be expressed directly in terms of the determinants of the autocorrelation matrices $\mathbf{R}_{N+1}$ and $\mathbf{R}_N$. This result is of considerable value in theoretical analysis.

Consider the augmented normal equation Eq. (2.20) for the Nth-order predictor. We can further augment this equation to include the information about the lower-order predictors by appending more *columns*. To demonstrate, let $N = 3$. We then obtain four sets of augmented normal equations, one for each order, which can be written elegantly together as follows:

$$
\mathbf{R}_4 \times
\begin{bmatrix}
1 & 0 & 0 & 0 \\
a_{3,1} & 1 & 0 & 0 \\
a_{3,2} & a_{2,1} & 1 & 0 \\
a_{3,3} & a_{2,2} & a_{1,1} & 1
\end{bmatrix}
=
\begin{bmatrix}
\mathcal{E}_3^f & \times & \times & \times \\
0 & \mathcal{E}_2^f & \times & \times \\
0 & 0 & \mathcal{E}_1^f & \times \\
0 & 0 & 0 & \mathcal{E}_0^f
\end{bmatrix}
\tag{2.37}
$$

This uses the fact that $\mathbf{R}_3$ is a submatrix of $\mathbf{R}_4$, that is,

$$
\mathbf{R}_4 =
\begin{bmatrix}
R(0) & \times \\
\times & \mathbf{R}_3
\end{bmatrix},
\tag{2.38}
$$

and similarly, $\mathbf{R}_2$ is a submatrix of $\mathbf{R}_3$ and so forth. The entries $\times$ are possibly nonzero but will not enter our discussion. Taking determinants, we arrive at

$$\det \mathbf{R}_4 = \mathcal{E}_3^f \mathcal{E}_2^f \mathcal{E}_1^f \mathcal{E}_0^f,$$

where we have used the fact that the determinant of a triangular matrix is equal to the product of its diagonal elements (Horn and Johnson, 1985). Extending this for arbitrary N, we have the following result:

$$\det \mathbf{R}_{N+1} = \mathcal{E}_N^f \mathcal{E}_{N-1}^f \, \cdots \, \mathcal{E}_0^f. \tag{2.39}$$

In a similar manner, we have

$$\det \mathbf{R}_N = \mathcal{E}_{N-1}^f \mathcal{E}_{N-2}^f \, \cdots \, \mathcal{E}_0^f. \tag{2.40}$$

Taking the ratio, we arrive at

$$\mathcal{E}_N^f = \frac{\det \mathbf{R}_{N+1}}{\det \mathbf{R}_N}. \tag{2.41}$$

Thus, the minimized mean square prediction errors can be expressed directly in terms of the determinants of two autocorrelation matrices. In Section 6.5.2, we will further show that

$$\lim_{N \to \infty} \left(\det \mathbf{R}_N \right)^{1/N} = \lim_{N \to \infty} \mathcal{E}_N^f \tag{2.42}$$

That is, the limiting value of the mean squared prediction error is identical to the limiting value of $\left(\det \mathbf{R}_N \right)^{1/N}$.

2.5   ESTIMATING THE AUTOCORRELATION

In any practical application that involves linear prediction, the autocorrelation samples $R(k)$ have to be estimated from (possibly noisy) measured data $x(n)$ representing the random process. There are many methods for this, two of which are described here.

The autocorrelation method. In this method, we define the truncated version of measured data

$$x_L(n) = \begin{cases} x(n) & 0 \le n \le L-1, \\ 0 & \text{outside,} \end{cases}$$

and compute its deterministic autocorrelation. Thus, the estimate of $R(k)$ has the form

$$\widehat{R}(k) = \sum_{n=0}^{L-1} x_L(n) x_L^* (n - k). \tag{2.43}$$

There are variations of this method; for example, one could divide the summation by the number of terms in the sum (which depends on k) and so forth. From the estimated value $\widehat{R}(k)$, we form the Toeplitz matrix $\mathbf{R}_N$ and proceed with the computation of the predictor coefficients.

A simple matrix interpretation of the estimation of $\mathbf{R}_N$ is useful. For example, if we have $L = 5$ and wish to estimate the 3×3 autocorrelation matrix $\mathbf{R}_3$, the computation (2.43) is equivalent to defining the data matrix

$$\mathbf{X} = \begin{bmatrix} x(0) & 0 & 0 \\ x(1) & x(0) & 0 \\ x(2) & x(1) & x(0) \\ x(3) & x(2) & x(1) \\ x(4) & x(3) & x(2) \\ 0 & x(4) & x(3) \\ 0 & 0 & x(4) \end{bmatrix} \tag{2.44}$$

and forming the estimate of $\mathbf{R}_3$ using

$$\widehat{\mathbf{R}}_3 = (\mathbf{X}^\dagger \mathbf{X})^*. \tag{2.45}$$

This is a Toeplitz matrix whose top row is $\widehat{R}(0), \widehat{R}(1), \ldots$. Furthermore, it is positive definite by virtue of its form $\mathbf{X}^\dagger \mathbf{X}$. So, all the properties from linear prediction theory continue to be satisfied by the polynomial $A_N(z)$. For example, we can use Levinson's recursion (Section 3.2) to solve for $A_N(z)$, and all zeros of $A_N(z)$ are guaranteed to be in $|z| < 1$ (Appendix C).

Note that the number of samples of $x(n)$ used in the estimate of $R(k)$ decreases as k increases, so the estimates are of good quality only if k is small compared with the number of available data samples L. There are variations of this method that use a tapering window on the given data instead of abruptly truncating it (Rabiner and Schafer, 1978; Therrien, 1992).

The covariance method. In a variation called the "covariance method," the data matrix $\mathbf{X}$ is formed differently:

$$\mathbf{X} = \begin{bmatrix} x(2) & x(1) & x(0) \\ x(3) & x(2) & x(1) \\ x(4) & x(3) & x(2) \end{bmatrix} \tag{2.46}$$

and the autocorrelation estimated as $\widehat{\mathbf{R}}_3 = (\mathbf{X}^\dagger \mathbf{X})^*$. If necessary we can divide each element of the estimate by a fixed integer so that this looks like a time average.

In the covariance method, each $\widehat{R}(k)$ is an average of N possibly nonzero samples of data, unlike the autocorrelation method, where the number of samples used in the estimate of $R(k)$ decreases as k increases. So, the estimates, in general, tend to be better than in the autocorrelation method, but the matrix $\widehat{\mathbf{R}}_N$ in this case is not necessarily Toeplitz. So, Levinson's recursion (Section 3.2) cannot be applied for solving the normal equations, and we have to solve the normal equations directly. Another problem with non-Toeplitz covariances is that the solution $A_N(z)$ is not guaranteed to have all zeros inside the unit circle.

More detailed discussions on the relative advantages and disadvantages of these methods can be found in many references (e.g., Makhoul, 1975; Rabiner and Schafer, 1978; Kay, 1988; Therrien, 1992).

2.6   CONCLUDING REMARKS

In this chapter, we introduced the linear prediction problem and discussed its solution. The solution appears in the form of a set of linear equations called the normal equations. An efficient way to solve the normal equations using a recursive procedure, called Levinson's recursion, will be introduced in the next chapter. This recursion will place in evidence a structure called the *lattice structure* for linear prediction. Deeper discussions on linear prediction will follow in later chapters. The extension of the optimal prediction problem for the case of vector processes, the MIMO LPC problem, will be considered in Chapter 8.

• • • •

CHAPTER 3

Levinson's Recursion

3.1 INTRODUCTION

The optimal linear predictor coefficients $a_{N,i}$ are solutions to the set of normal equations given by Eq. (2.8). Traditional techniques to solve these equations require computations of the order of N^3 (see, e.g., Golub and Van Loan, 1989). However, the $N \times N$ matrix $\mathbf{R}_N$ in these equations is not arbitrary, but Toeplitz. This property can be exploited to solve these equations in an efficient manner, requiring computations of the order of N^2. A recursive procedure for this, due to Levinson (1947), will be described in this chapter. This procedure, in addition to being efficient, also places in evidence many useful properties of the optimal predictor, as we shall see in the next several sections.

3.2 DERIVATION OF LEVINSON'S RECURSION

Levinson's recursion is based on the observation that if the solution to the predictor problem is known for order m, then the solution for order $(m+1)$ can be obtained by a simple updating process. In this way, we obtain not only the solution to the Nth-order problem, but also all the lower orders. To demonstrate the basic idea, consider the third-order prediction problem. The augmented normal Eq. (2.20) becomes

$$\underbrace{\begin{bmatrix} R(0) & R(1) & R(2) & R(3) \\ R^*(1) & R(0) & R(1) & R(2) \\ R^*(2) & R^*(1) & R(0) & R(1) \\ R^*(3) & R^*(2) & R^*(1) & R(0) \end{bmatrix}}_{\mathbf{R}_4} \underbrace{\begin{bmatrix} 1 \\ a_{3,1} \\ a_{3,2} \\ a_{3,3} \end{bmatrix}}_{\mathbf{a}_3} = \begin{bmatrix} \mathcal{E}_3^f \\ 0 \\ 0 \\ 0 \end{bmatrix}. \qquad (3.1)$$

This set of four equations describes the third-order optimal predictor. Our aim is to show how we can pass from the third-order to the fourth-order case. For this, note that we can append a fifth equation to the above set of four and write it as

$$\underbrace{\begin{bmatrix} R(0) & R(1) & R(2) & R(3) & R(4) \\ R^*(1) & R(0) & R(1) & R(2) & R(3) \\ R^*(2) & R^*(1) & R(0) & R(1) & R(2) \\ R^*(3) & R^*(2) & R^*(1) & R(0) & R(1) \\ R^*(4) & R^*(3) & R^*(2) & R^*(1) & R(0) \end{bmatrix}}_{\mathbf{R}_5} \begin{bmatrix} 1 \\ a_{3,1} \\ a_{3,2} \\ a_{3,3} \\ 0 \end{bmatrix} = \begin{bmatrix} \mathcal{E}_3^f \\ 0 \\ 0 \\ 0 \\ \alpha_3 \end{bmatrix}. \tag{3.2}$$

where

$$\alpha_3 \overset{\Delta}{=} R^*(4) + a_{3,1}R^*(3) + a_{3,2}R^*(2) + a_{3,3}R^*(1). \tag{3.3}$$

The matrix $\mathbf{R}_5$ above is Hermitian and Toeplitz. Using this, we verify that its elements satisfy (Problem 9)

$$R_{4-i,4-k}^* = R_{ik}, \quad 0 \le i, k \le 4. \tag{3.4}$$

In other words, if we reverse the order of all rows, then reverse the order of all columns and then conjugate the elements, the result is the same matrix! As a result, Eq. (3.2) also implies

$$\underbrace{\begin{bmatrix} R(0) & R(1) & R(2) & R(3) & R(4) \\ R^*(1) & R(0) & R(1) & R(2) & R(3) \\ R^*(2) & R^*(1) & R(0) & R(1) & R(2) \\ R^*(3) & R^*(2) & R^*(1) & R(0) & R(1) \\ R^*(4) & R^*(3) & R^*(2) & R^*(1) & R(0) \end{bmatrix}}_{\mathbf{R}_5} \begin{bmatrix} 0 \\ a_{3,3}^* \\ a_{3,2}^* \\ a_{3,1}^* \\ 1 \end{bmatrix} = \begin{bmatrix} \alpha_3^* \\ 0 \\ 0 \\ 0 \\ \mathcal{E}_3^f \end{bmatrix}. \tag{3.5}$$

If we now take a linear combination of Eqs. (3.2) and (3.5) such that the last element on the right-hand side becomes zero, we will obtain the equations governing the fourth-order predictor indeed! Thus, consider the operation

$$\text{Eq. (3.2)} \quad + \quad k_4^* \times \text{Eq. (3.5)}, \tag{3.6}$$

where k_4 is a constant. If we choose

$$k_4^* = \frac{-\alpha_3}{\mathcal{E}_3^f}, \tag{3.7}$$

then the result has the form

$$
\underbrace{\begin{bmatrix} R(0) & R(1) & R(2) & R(3) & R(4) \\ R^*(1) & R(0) & R(1) & R(2) & R(3) \\ R^*(2) & R^*(1) & R(0) & R(1) & R(2) \\ R^*(3) & R^*(2) & R^*(1) & R(0) & R(1) \\ R^*(4) & R^*(3) & R^*(2) & R^*(1) & R(0) \end{bmatrix}}_{\mathbf{R}_5} \begin{bmatrix} 1 \\ a_{4,1} \\ a_{4,2} \\ a_{4,3} \\ a_{4,4} \end{bmatrix} = \begin{bmatrix} \times \\ 0 \\ 0 \\ 0 \\ 0 \end{bmatrix}, \tag{3.8}
$$

where

$$
a_{4,1} = a_{3,1} + k_4^* \, a_{3,3}^*
$$
$$
a_{4,2} = a_{3,2} + k_4^* \, a_{3,2}^*
$$
$$
a_{4,3} = a_{3,3} + k_4^* \, a_{3,1}^*
$$
$$
a_{4,4} = k_4^* .
$$

By comparison with Eq. (3.1), we conclude that the element denoted $\times$ on the right-hand side of Eq. (3.8) is $\mathcal{E}_4^f$, which is the minimized forward prediction error for the fourth-order predictor. From the above construction, we see that this is related to $\mathcal{E}_3^f$ as

$$
\begin{aligned}
\mathcal{E}_4^f &= \mathcal{E}_3^f + k_4^* \, \alpha_3^* \\
&= \mathcal{E}_3^f - k_4 k_4^* \, \mathcal{E}_3^f \quad \text{(from Eq. (3.7))}. \\
&= (1 - |k_4|^2) \mathcal{E}_3^f.
\end{aligned}
$$

Summarizing, if we know the coefficients $a_{3,i}$ and the mean square error $\mathcal{E}_3^f$ for the third-order optimal predictor, we can find the corresponding quantities for the fourth-order predictor from the above equations.

Polynomial Notation. The preceding computation of $\{a_{4,k}\}$ from $\{a_{3,k}\}$ can be written more compactly if we use polynomial notations and define the FIR filter

$$
A_m(z) = 1 + a_{m,1}^* \, z^{-1} + a_{m,2}^* \, z^{-2} + \ldots + a_{m,m}^* \, z^{-m} \quad \text{(prediction polynomial)}. \tag{3.9}
$$

With this notation, we can rewrite the computation of $a_{4,k}$ from $a_{3,k}$ as

$$
A_4(z) = A_3(z) + k_4 z^{-1} [z^{-3} \widetilde{A}_3(z)], \tag{3.10}
$$

where the tilde notation is as defined in Section 1.2. Thus,

$$
z^{-m} \widetilde{A}_m(z) = a_{m,m} + a_{m,m-1} z^{-1} + \ldots + a_{m,1} z^{-(m-1)} + z^{-m}. \tag{3.11}
$$

3.2.1   Summary of Levinson's Recursion

The recursion demonstrated above for the third-order case can be generalized easily to arbitrary predictor orders. Thus, let $A_m(z)$ be the predictor polynomial for the mth-order optimal predictor and $\mathcal{E}_m^f$ the corresponding mean square value of the prediction error. Then, we can find the corresponding quantities for the $(m+1)$th-order optimal predictor as follows:

$$k_{m+1} = \frac{-\alpha_m^*}{\mathcal{E}_m^f}, \qquad (3.12)$$

$$A_{m+1}(z) = A_m(z) + k_{m+1}z^{-1}[z^{-m}\tilde{A}_m(z)] \quad \text{(order update)}, \qquad (3.13)$$

$$\mathcal{E}_{m+1}^f = (1 - |k_{m+1}|^2)\mathcal{E}_m^f, \quad \text{(error update)}, \qquad (3.14)$$

where

$$\alpha_m = R^*(m+1) + a_{m,1}R^*(m) + a_{m,2}R^*(m-1) + \ldots + a_{m,m}R^*(1). \qquad (3.15)$$

Initialization. Once this recursion is initialized for small m (e.g., $m = 0$), it can be used to solve the optimal predictor problem for any m. Note that for $m = 0$, the predictor polynomial is

$$A_0(z) = 1, \qquad (3.16)$$

and the error is, from Eq. (2.3), $e_0^f(n) = x(n)$. Thus,

$$\mathcal{E}_0^f = R(0). \qquad (3.17)$$

Also, from the definition of α_m, we have

$$\alpha_0 = R^*(1). \qquad (3.18)$$

The above three equations are used to initialize Levinson's recursion. For example, we can compute

$$k_1 = -\alpha_0^* / \mathcal{E}_0^f = -R(1)/R(0), \qquad (3.19)$$

and evaluate $A_1(z)$ from Eq. (3.13) and so forth. $\square$

3.2.2   The Partial Correlation Coefficient

The quantity k_i in Levinson's recursion has a nice "physical" significance. Recall from Section 2.3 that the error $e_m^f(n)$ of the optimal predictor is orthogonal to the past samples

$$x(n-1), x(n-2), \ldots, x(n-m). \qquad (3.20)$$

The "next older" sample $x(n - m - 1)$, however, is not necessarily orthogonal to the error, that is, the correlation $E[e_m^f(n)x^*(n - m - 1)]$ could be nonzero. It can be shown (Problem 6) that the quantity α_m^* in the numerator of k_{m+1} satisfies the relation

$$\alpha_m^* = E[e_m^f(n)x^*(n - m - 1)]. \qquad (3.21)$$

The coefficient k_{m+1} represents this correlation, normalized by the mean square error $\mathcal{E}_m^f$. The correction to the prediction polynomial in the order-update equation (second term in Eq. (3.13)) is proportional to this normalized correlation.

In the literature, the coefficients k_i have been called the *partial correlation* coefficients and abbreviated as *parcor* coefficients. They are also known as the *lattice* or *reflection* coefficients, for reasons that will become clear in later chapters.

Example 3.1: Levinson's Recursion. Consider again the real WSS process with the autocorrelation (2.12). We initialize Levinson's recursion as described above, that is,

$$A_0(z) = 1, \quad \alpha_0 = R(1) = 1.5, \quad \text{and} \quad \mathcal{E}_0^f = R(0) = 2.1. \qquad (3.22)$$

We can now compute the quantities

$$k_1 = -\alpha_0/\mathcal{E}_0^f = -5/7$$
$$A_1(z) = A_0(z) + k_1 z^{-1}\widetilde{A}_0(z) = 1 - (5/7)z^{-1},$$
$$\mathcal{E}_1^f = (1 - k_1^2)\mathcal{E}_0^f = 36/35.$$

We have now obtained all information about the first-order predictor. To obtain the second-order predictor, we compute the appropriate quantities in the following order:

$$\alpha_1 = R(2) + a_{1,1}R(1) = (9/10) - (5/7) \times 1.5 = -6/35$$
$$k_2 = -\alpha_1/\mathcal{E}_1^f = 1/6$$
$$A_2(z) = A_1(z) + k_2 z^{-2}\widetilde{A}_1(z) = 1 - (5/6)z^{-1} + (1/6)z^{-2}$$
$$\mathcal{E}_2^f = (1 - k_2^2)\mathcal{E}_1^f = 1.0.$$

We can proceed in this way to compute optimal predictors of any order. The above results agree with those of Example 2.1, where we used a direct approach to solve the normal equations.

3.3   SIMPLE PROPERTIES OF LEVINSON'S RECURSION

Inspection of Levinson's recursion reveals many interesting properties. Some of these are discussed below. Deeper properties will be studied in the next several sections.

1. *Computational complexity.* The computation of $A_{m+1}(z)$ from $A_m(z)$ requires nearly m multiplication and addition operations. As a result, the amount of computation required

for N repetitions (i.e., to find $a_{N,i}$) is proportional to N^2. This should be compared with traditional methods for solving Eq. (2.8) (such as Gaussian elimination), which require computations proportional to N^3. In addition to reducing the computation, Levinson's recursion also reveals the optimal solution to *all the predictors* of orders $\leq N$.

2. *Parcor coefficients are bounded.* Because $\mathcal{E}_m^f$ is the mean square value of the prediction error, we know $\mathcal{E}_m^f \geq 0$ for any m. From Eq. (3.14), it, therefore, follows that $|k_i|^2 \leq 1$, for any i.

3. *Strict bound on parcor coefficients.* Let $|k_{m+1}| = 1$. Then Eq. (3.14) says that $\mathcal{E}_{m+1}^f = 0$. In other words, the mean square value of the prediction error sequence $e_{m+1}^f(n)$ is zero. This means that $e_{m+1}^f(n)$ is identically zero for all n. In other words, the output of $A_{m+1}(z)$ in response to $x(n)$ is zero (Fig. 2.1(a)). Using an argument similar to the one in Section 2.4.2, we conclude that this is not possible unless $x(n)$ is a line spectral process. Summarizing, we have

$$|k_i| < 1, \quad \text{for any } i, \tag{3.23}$$

unless $x(n)$ is a line spectral process. Recall also that the assumption that $x(n)$ is not line spectral also ensures that the autocorrelation matrix $\mathbf{R}_N$ in the normal equations is nonsingular for all N (Section 2.4.2).

4. *Minimized error is monotone.* From Eq. (3.14) it also follows that the mean square error is a monotone function, that is,

$$\mathcal{E}_{m+1}^f \leq \mathcal{E}_m^f, \tag{3.24}$$

with equality if and only if $k_{m+1} = 0$. It is interesting to note that repeated application of Eq. (3.14) yields the following expression for the mean square prediction error of the Nth-order optimal predictor:

$$\begin{aligned}
\mathcal{E}_N^f &= \left(1 - |k_N|^2\right)\left(1 - |k_{N-1}|^2\right)\cdots\left(1 - |k_1|^2\right)\mathcal{E}_0^f \\
&= \left(1 - |k_N|^2\right)\left(1 - |k_{N-1}|^2\right)\cdots\left(1 - |k_1|^2\right)R(0)
\end{aligned}$$

Example 3.2: Levinson's recursion. Now, consider a WSS process $x(n)$ with power spectrum $S_{xx}(e^{j\omega})$ as shown in Fig. 3.1 (top). The first few samples of the autocorrelation are

$R(0)$	$R(1)$	$R(2)$	$R(3)$	$R(4)$	$R(5)$	$R(6)$
0.1482	0.0500	0.0170	−0.0323	−0.0629	0.0035	−0.0087

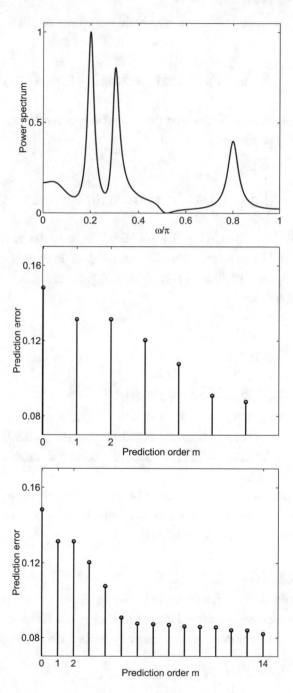

FIGURE 3.1: Example 3.2. The input power spectrum $S_{xx}(e^{j\omega})$ (top), the prediction error $\mathcal{E}_m^f$ plotted for a few prediction orders m (middle), and the prediction error $\mathcal{E}_m^f$ shown for more orders (bottom).

If we perform Levinson's recursion with this, we obtain the parcor coefficients

k_1	k_2	k_3	k_4	k_5	k_6
-0.3374	-0.0010	0.2901	0.3253	-0.3939	0.1908

for the first six optimal predictors. Notice that these coefficients satisfy $|k_m| < 1$ as expected from theory. The corresponding prediction errors are

$\mathcal{E}_0^f$	$\mathcal{E}_1^f$	$\mathcal{E}_2^f$	$\mathcal{E}_3^f$	$\mathcal{E}_4^f$	$\mathcal{E}_5^f$	$\mathcal{E}_6^f$
0.1482	0.1314	0.1314	0.1203	0.1076	0.0909	0.0876

The prediction errors are also plotted in Fig. 3.1 (middle). The error decreases monotonically as the prediction order increases. The same trend continues for higher prediction orders as demonstrated in the bottom plot. The optimal prediction polynomials $A_0(z)$ through $A_6(z)$ have the coefficients given in the following table:

$A_0(z)$	1.0					
$A_1(z)$	1.0	-0.3374				
$A_2(z)$	1.0	-0.3370	-0.0010			
$A_3(z)$	1.0	-0.3373	-0.0988	0.2901		
$A_4(z)$	1.0	-0.2429	-0.1309	0.1804	0.3253	
$A_5(z)$	1.0	-0.3711	-0.2020	0.2320	0.4210	-0.3939
$A_6(z)$	1.0	-0.4462	-0.1217	0.2762	0.3825	-0.4647	0.1908

The zeros of these polynomials can be verified to be inside the unit circle, as expected from our theoretical development (Appendix C). For example, the zeros of $A_6(z)$ are complex conjugate pairs with magnitudes 0.9115, 0.7853, and 0.6102.

3.4   THE WHITENING EFFECT

Suppose we compute optimal linear predictors of increasing order using Levinson's recursion. We know the prediction error decreases with order, that is, $\mathcal{E}_{k+1} \leq \mathcal{E}_k$. Assume that after the predictor has reached the order m, the error does not decrease any further, that is, suppose that

$$\mathcal{E}_m^f = \mathcal{E}_{m+1}^f = \mathcal{E}_{m+2}^f = \mathcal{E}_{m+3}^f = \cdots \qquad (3.25)$$

This represents a *stalling condition*, that is, increasing the number of past samples does not help to increase the prediction accuracy any further. Whenever such a stalling occurs, it turns out that the error $e_m^f(n)$ satisfies a very interesting property, namely, any pair of samples are mutually orthogonal. That is,

$$E\left[e_m^f(n)[e_m^f(n+i)]^*\right] = \begin{cases} 0 & i \neq 0 \\ \mathcal{E}_m^f & i = 0. \end{cases} \qquad (3.26)$$

In particular, if $x(n)$ has zero mean, then $e_m^f(n)$ will also have zero mean, and the preceding equation means that $e_m^f(n)$ is *white*. Now, from Section 2.2, we know that $x(n)$ can be represented as the output of a filter, excited with $e_m^f(n)$ (see Fig. 2.1(b)). Thus, whenever stalling occurs, $x(n)$ is the output of an all-pole filter $1/A_m(z)$ with *white* input; such a zero-mean process $x(n)$ is said to be *autoregressive* (AR). We will present a more complete discussion of AR processes in Section 5.2. For the case of nonzero mean, Eq. (3.26) still holds, and we say that $e_m^f(n)$ is an orthogonal (rather than white) random process.

Theorem 3.1. *Stalling and whitening.* Consider the optimal prediction of a WSS process $x(n)$, with successively increasing predictor orders. Then, the iteration stalls after the mth-order predictor (i.e., the mean square error stays constant as shown by Eq. (3.25)), if and only if the prediction error $e_m^f(n)$ satisfies the orthogonality condition Eq. (3.26). $\diamond$

Proof. Stalling implies, in particular, that $\mathcal{E}_{m+1}^f = \mathcal{E}_m^f$, that is, $k_{m+1} = 0$ (from Eq. (3.14)). As a result, $A_{m+1}(z) = A_m(z)$. Repeating this, we see that

$$A_m(z) = A_{m+1}(z) = A_{m+2}(z) = \dots \qquad (3.27)$$

The prediction error sequence $e_\ell^f(n)$, therefore, is the same for all $\ell \geq m$. The condition $k_{m+1} = 0$ implies $\alpha_m = 0$ from Eq. (3.12). In other words, the cross-correlation Eq. (3.21) is zero. Repeating this argument, we see that whenever stalling occurs, we have

$$E[e_{m+\ell}^f(n)x^*(n-m-\ell-1)] = 0, \quad \ell \geq 0. \qquad (3.28)$$

But $e_{m+\ell}^f(n) = e_m^f(n)$ for any $\ell \geq 0$, so that

$$E[e_m^f(n)x^*(n-m-\ell-1)] = 0, \quad \ell \geq 0. \qquad (3.29)$$

By orthogonality principle, we already know that $E[e_m^f(n)x^*(n-i)] = 0$ for $1 \leq i \leq m$. Combining the preceding two equations, we conclude that

$$E[e_m^f(n)x^*(n-\ell)] = 0, \quad \ell \geq 1. \qquad (3.30)$$

In other words, the error $e_m^f(n)$ is orthogonal to all the past samples of $x(n)$. We also know that $e_m^f(n)$ is a linear combination of present and past samples of $x(n)$, that is,

$$e_m^f(n) = x(n) + \sum_{i=1}^{m} a_{m,i}^* \, x(n-i). \tag{3.31}$$

Similarly,

$$e_m^f(n-\ell) = x(n-\ell) + \sum_{i=1}^{m} a_{m,i}^* \, x(n-\ell-i). \tag{3.32}$$

Because $e_m^f(n)$ is orthogonal to all the past samples of $x(n)$, we, therefore, conclude that $e_m^f(n)$ is orthogonal to $e_m^f(n-\ell)$, $\ell > 0$. Summarizing, we have proved, $E[e_m^f(n)[e_m^f(n-\ell)]^*] = 0$ for $\ell > 0$. This proves Eq. (3.26) indeed. By reversing the above argument, we can show that if $e_m^f(n)$ has the property (3.26) then the recursion stalls (i.e., Eq. (3.25) holds).   □

Example 3.3: Stalling and Whitening. Consider a real WSS process with autocorrelation

$$R(k) = \rho^{|k|}, \tag{3.33}$$

where $-1 < \rho < 1$. The first-order predictor coefficient $a_{1,1}$ is obtained by solving

$$R(0)a_{1,1} = -R(1), \tag{3.34}$$

so that $a_{1,1} = -\rho$. Thus, the optimal predictor polynomial is $A_1(z) = 1 - \rho z^{-1}$. To compute the second-order predictor, we first evaluate

$$\alpha_1 = R(2) + a_{1,1}R(1) = \rho^2 - \rho^2 = 0. \tag{3.35}$$

Using Levinson's recursion, we find $k_2 = -\alpha_1^*/\mathcal{E}_1^f = 0$, so that $A_2(z) = A_1(z)$. To find the third-order predictor, note that

$$\alpha_2 = R(3) + a_{2,1}R(2) + a_{2,2}R(1) = R(3) - \rho R(2) = 0. \tag{3.36}$$

Thus, $k_3 = 0$ and $A_3(z) = A_1(z)$. No matter how far we continue, we will find in this case that $A_m(z) = A_1(z)$ for all m. That is, the recursion has stalled. Let us double-check this by testing whether $e_1^f(n)$ satisfies Eq. (3.26). Because $e_1^f(n)$ is the output of $A_1(z)$ in response to $x(n)$, we have

$$e_1^f(n) = x(n) - \rho x(n-1). \tag{3.37}$$

Thus, for $k > 0$,

$$E[e_1^f(n)e_1^f(n-k)] = R(k) + \rho^2 R(k) - \rho R(k-1) - \rho R(k+1) = 0, \qquad (3.38)$$

as anticipated. Because $R(\infty) = 0$, the process $x(n)$ has zero mean. So, $e_1^f(n)$ is a zero-mean white process.

3.5   CONCLUDING REMARKS

Levinson's work was done in 1947 and, in his own words, was a "mathematically trivial procedure." However, it is clear from this chapter that Levinson's recursion is very elegant and insightful. It can be related to early work by other famous authors (e.g., Chandrasekhar, 1947). It is also related to the Berlekamp–Massey algorithm (Berlekamp, 1969). For a fascinating history, the reader should study the scholarly review by Kaliath (1974; in particular, see p. 160). The derivation of Levinson's recursion in this chapter used the properties of the autocorrelation matrix. However, the method can be extended to the case of Toeplitz matrices, which are not necessarily positive definite (Blahut, 1985).

In fact, even the Toeplitz structure is not necessary if the goal is to obtain an $O(N^2)$ algorithm. In 1979, Kailath et al. introduced the idea of *displacement rank* for matrices. They showed that as long as the displacement rank is a fixed number independent of matrix size, $O(N^2)$ algorithms can be found for solving linear equations involving these matrices. It turns out that Toeplitz matrices have displacement rank 2, regardless of the matrix size. The same is true for inverses of Toeplitz matrices (which are not necessarily Toeplitz).

One of the outcomes of Levinson's recursion is that it gives rise to an elegant structure for linear prediction called the *lattice structure*. This will be the topic of discussion for the next chapter.

· · · ·

CHAPTER 4

Lattice Structures for Linear Prediction

4.1 INTRODUCTION

In this chapter, we present lattice structures for linear prediction. These structures essentially follow from Levinson's recursion. Lattice structures have fundamental importance not only in linear prediction theory but, more generally, in signal processing. For example, they arise in the theory of *all-pass filters:* any stable rational all-pass filter can be represented using an IIR lattice structure similar to the IIR LPC lattice. As a preparation for the main topic, we first discuss the idea of backward linear prediction.

4.2 THE BACKWARD PREDICTOR

Let $x(n)$ represent a WSS random process as usual. A backward linear predictor for this process estimates the sample $x(n - N - 1)$ based on the "future values" $x(n - 1), \ldots, x(n - N)$. The predicted value is a linear combination of the form

$$\widehat{x}_N^{\text{b}}(n - N - 1) = -\sum_{i=1}^{N} b_{N,i}^* x(n - i), \qquad (4.1)$$

and the prediction error is

$$
\begin{aligned}
e_N^{\text{b}}(n) &= x(n - N - 1) - \widehat{x}_N^{\text{b}}(n - N - 1) \\
&= \sum_{i=1}^{N} b_{N,i}^* x(n - i) + x(n - N - 1).
\end{aligned}
$$

The superscript b is meant to be a reminder of "backward." Figure 4.1 demonstrates the difference between the forward and backward predictors. Notice that both $\widehat{x}_N^{f}(n)$ and $\widehat{x}_N^{\text{b}}(n - N - 1)$ are based on the same set of N measurements. The backward prediction problem is primarily of theoretical interest, but it helps us to attach a "physical" significance to the polynomial $z^{-m}\widetilde{A}_m(z)$ in Levinson's

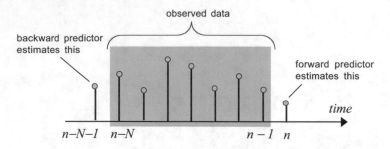

FIGURE 4.1: Comparison of the forward and backward predictors.

recursion (3.13), as we shall see. As in the forward predictor problem, we again define the predictor polynomial

$$B_N(z) = \sum_{i=1}^{N} b_{N,i}^* z^{-i} + z^{-(N+1)}. \tag{4.2}$$

The output of this FIR filter in response to $x(n)$ is equal to the predictor error $e_N^b(n)$.

To find the optimal predictor coefficients $b_{N,i}^*$, we again apply the orthogonality principle, which says that $e_N^b(n)$ must be orthogonal to $x(n - k)$, for $1 \le k \le N$. Using this, we can show that the solution is closely related to that of the forward predictor of Section 2.3. With $a_{N,i}$ denoting the optimal forward predictor coefficients, it can be shown (Problem 15) that

$$b_{N,i} = a_{N,N+1-i}^*, \quad 1 \le i \le N, \tag{4.3}$$

This equation represents time-reversal and conjugation. Thus, the optimum backward predictor polynomial is given by

$$B_N(z) = z^{-(N+1)} \widetilde{A}_N(z), \tag{4.4}$$

where the tilde notation is as defined in Section 1.2.1. Figure 4.2 summarizes how the forward and backward prediction errors are derived from $x(n)$ by using the two FIR filters, $A_N(z)$ and $B_N(z)$. It is therefore clear that we can derive the backward prediction error sequence $e_N^b(n)$ from the forward prediction error sequence $e_N^f(n)$ as shown in Fig. 4.3. In view of Eq. (4.4), we have

$$\frac{B_N(z)}{A_N(z)} = \frac{z^{-(N+1)} \widetilde{A}_N(z)}{A_N(z)} \tag{4.5}$$

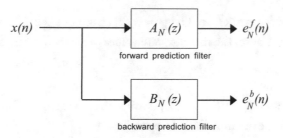

FIGURE 4.2: The forward and the backward prediction filters.

4.2.1 All-Pass Property

We now show that the function

$$G_N(z) \triangleq \frac{z^{-N}\widetilde{A}_N(z)}{A_N(z)} \tag{4.6}$$

is all-pass, that is,

$$|G_N(e^{j\omega})| = 1, \quad \forall\, \omega. \tag{4.7}$$

To see this, simply observe that

$$\widetilde{G}_N(z)\, G_N(z) = \frac{z^N A_N(z)}{\widetilde{A}_N(z)} \times \frac{z^{-N}\widetilde{A}_N(z)}{A_N(z)} = 1, \quad \forall\, z$$

But because

$$\widetilde{H}_N(e^{j\omega}) = H_N^*(e^{j\omega}),$$

the preceding implies that $|G_N(e^{j\omega})|^2 = 1$, which proves Eq. (4.7). From Fig. 4.3, we therefore see that the backward error $e_N^b(n)$ is the output of the all-pass filter $z^{-1}G_N(z)$ in response to the

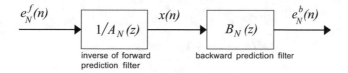

FIGURE 4.3: The backward prediction error sequence, derived from the forward prediction error sequence.

input $e_N^f(n)$, which is the forward error. The power spectrum of $e_N^f(n)$ is therefore identical to that of $e_N^b(n)$. In particular, therefore, the mean square values of the two errors are identical, that is,

$$\mathcal{E}_N^b = \mathcal{E}_N^f \qquad (4.8)$$

where $\mathcal{E}_N^b = E[|e_N^b(n)|^2]$ and $\mathcal{E}_N^f = E[|e_N^f(n)|^2]$. Notice that the all-pass filter (4.6) is stable because the zeros of $A_N(z)$ are inside the unit circle (Appendix C).

4.2.2   Orthogonality of the Optimal Prediction Errors

Consider the set of backward prediction error sequences of various orders, that is,

$$e_0^b(n), \ e_1^b(n), \ e_2^b(n), \dots \qquad (4.9)$$

We will show that any two of these are orthogonal. More precisely,

Theorem 4.1. *Orthogonality of errors.* The backward prediction error sequences satisfy the property

$$E\left[e_m^b(n)[e_k^b(n)]^*\right] = \begin{cases} 0 & \text{for } k \neq m \\ \mathcal{E}_m^b & \text{for } k = m. \end{cases} \qquad (4.10)$$

for all $k, m \geq 0$. ◇

Proof. It is sufficient to prove this for $m > k$. According to the orthogonality principle, $e_m^b(n)$ is orthogonal to

$$x(n-1), \dots, x(n-m). \qquad (4.11)$$

From its definition,

$$e_k^b(n) = x(n-k-1) + b_{k,1}^* x(n-1) + b_{k,2}^* x(n-2) + \ \dots \ + b_{k,k}^* x(n-k).$$

From these two equations, we conclude that $E[e_m^b(n)[e_k^b(n)]^*] = 0$, for $m > k$. From this result, Eq. (4.10) follows immediately. □

In a similar manner, it can be shown (Problem 16) that the *forward* predictor error sequences have the following orthogonality property:

$$E\left[e_m^f(n)[e_k^f(n-1)]^*\right] = 0, \quad m > k. \qquad (4.12)$$

The above orthogonality properties have applications in adaptive filtering, specifically in improving the convergence of adaptive filters (Satorius and Alexander, 1979; Haykin, 2002). The process of generating the above set of orthogonal signals from the random process $x(n)$ has also been interpreted as a kind of Gram–Schmidt orthogonalization (Haykin, 2002). These details are beyond scope of this chapter, but the above references provide further details and bibliography.

4.3   LATTICE STRUCTURES

In Section 3.2, we presented Levinson's recursion, which computes the optimal predictor polynomial $A_{m+1}(z)$ from $A_m(z)$ according to

$$A_{m+1}(z) = A_m(z) + k_{m+1}z^{-1}[z^{-m}\widetilde{A}_m(z)] \qquad (4.13)$$

On the other hand, we know from Section 4.2, that the optimal backward predictor polynomial is given by Eq. (4.4). We can therefore rewrite Levinson's order update equation as

$$A_{m+1}(z) = A_m(z) + k_{m+1}B_m(z). \qquad (4.14)$$

From this, we also obtain

$$B_{m+1}(z) = z^{-1}[k_{m+1}^* A_m(z) + B_m(z)], \qquad (4.15)$$

by using $B_{m+1}(z) = z^{-(m+2)}\widetilde{A}_{m+1}(z)$. The preceding relations can be schematically represented as in Fig. 4.4(a). Because $e_m^f(n)$ and $e_m^b(n)$, are the outputs of the filters $A_m(z)$ and $B_m(z)$ in response to the common input $x(n)$, the prediction error signals are related as shown in Fig. 4.4(b). By

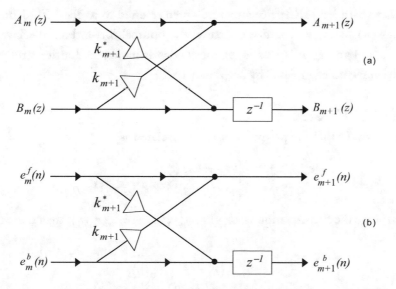

FIGURE 4.4: (a) The lattice section that generates the $(m + 1)$th-order predictor polynomials from the mth-order predictor polynomials. (b) The same system shown with error signals indicated as the inputs and outputs.

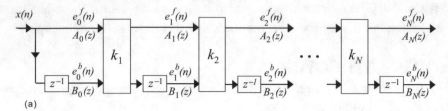

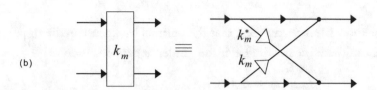

(b)

FIGURE 4.5: (a) The FIR LPC lattice structure that generates the prediction errors $e_m^f(n)$ and $e_m^b(n)$ for all optimal predictors with order $1 \leq m \leq N$. (b) Details of the mth lattice section.

using the facts that $A_0(z) = 1$ and $B_0(z) = z^{-1}$, we obtain the structure of Fig. 4.5(a). Here, each rectangular box is the two multiplier lattice section shown in Fig. 4.5(b). In this structure, the transfer functions $A_m(z)$ and $B_m(z)$ are generated by repeated use of the lattice sections in Fig. 4.4. This is a cascaded FIR lattice structure, often referred to as the LPC *FIR lattice*. If we apply the input $x(n)$ to the FIR lattice structure, the optimal forward and backward prediction error sequences $e_m^f(n)$ and $e_m^b(n)$ appear at various nodes as indicated. Lattice structures for linear prediction have been studied in detail by Makhoul (1977).

4.3.1 The IIR LPC Lattice

From Fig. 4.4(b), we see that the prediction errors are related as

$$e_{m+1}^f(n) = e_m^f(n) + k_{m+1}e_m^b(n),$$
$$e_{m+1}^b(n) = k_{m+1}^* e_m^f(n-1) + e_m^b(n-1).$$

Suppose we rearrange the first equation as $e_m^f(n) = e_{m+1}^f(n) - k_{m+1}e_m^b(n)$, then these two equations become

$$e_m^f(n) = e_{m+1}^f(n) - k_{m+1}e_m^b(n),$$
$$e_{m+1}^b(n) = k_{m+1}^* e_m^f(n-1) + e_m^b(n-1).$$

These equations have the structural representation shown in Fig. 4.6(a). Repeated application of this gives rise to the IIR structure of Fig. 4.6(b). This is precisely the IIR all-pass lattice structure

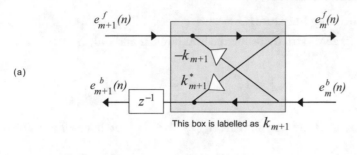

(a)

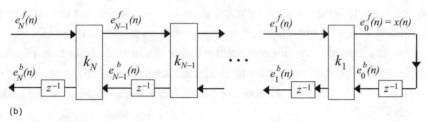

(b)

FIGURE 4.6: (a) The IIR LPC lattice section. (b) The IIR LPC lattice structure that generates the signal $x(n)$ from the optimal prediction error $e_N^f(n)$. The lower-order forward prediction errors $e_m^f(n)$ and the backward prediction errors are also generated in the process automatically.

studied in signal processing literature (Gray and Markel, 1973; Vaidyanathan, 1993; Oppenheim and Schafer, 1999). Thus, if we apply the forward prediction error $e_N^f(n)$ as an input to this structure, then the forward and backward error sequences of all the lower-order optimal predictors appear at various internal nodes, as indicated. In particular, the 0th-order prediction error $e_0^f(n)$ appears at the rightmost node. Because $e_0^f(n) = x(n)$, we therefore obtain the original random process $x(n)$ as an output of the IIR lattice, in response to the input $e_N^f(n)$. We already showed that the transfer function

$$z^{-1} G_N(z) = z^{-(N+1)} \frac{\widetilde{A}_N(z)}{A_N(z)}$$

is all-pass. This is the transfer function from the input $e_N^f(n)$ to the node $e_N^b(n)$ in the IIR lattice. The transfer function from the input to the rightmost node in Fig. 4.6(b) is the *all-pole* filter $1/A_N(z)$, because this transfer function produces $x(n)$ in response to the input $e_N^f(n)$ (see Fig. 4.2).

4.3.2 Stability of the IIR Filter

It is clear from these developments that the "parcor" coefficients k_m derived from Levinson's recursion are also the lattice coefficients for the all-pass function $z^{-N} \widetilde{A}_N(z)/A_N(z)$. Now, it is well known from the signal processing literature (Gray and Markel, 1973; Vaidyanathan, 1993) that the polynomial $A_N(z)$ has all its zeros strictly inside the unit circle if and only if

$$|k_m| < 1, \quad 1 \leq m \leq N. \tag{4.16}$$

In fact, under this condition, all the polynomials $A_m(z)$, $1 \leq m \leq N$ have all zeros strictly inside the unit circle.

On the other hand, Levinson's recursion (Section 3.2) has shown that the condition $|k_m| < 1$ is indeed true as long as $x(n)$ is not a line spectral process. This proves that the IIR filters $1/A_m(z)$ and $z^{-m}\widetilde{A}_m(z)/A_m(z)$ indeed have all poles strictly inside the unit circle so that Fig. 4.6(b) is, in particular, stable. Although the connection to the lattice structure is insightful, it is also possible to prove stability of $1/A_m(z)$ directly without the use of Levinson's recursion or the lattice. This is done in Appendix C.

4.3.3   The Upward and Downward Recursions

Levinson's recursion (Eq. (4.14) and (4.15)) computes the optimal predictor polynomials of increasing order, from a knowledge of the autocorrelation $R(k)$ of the WSS process $x(n)$. This is called the *upward* recursion. These equations can be inverted to obtain the relations

$$(1 - |k_{m+1}|^2) \times A_m(z) = A_{m+1}(z) - k_{m+1} z B_{m+1}(z),$$

$$(1 - |k_{m+1}|^2) \times B_m(z) = -k^*_{m+1} A_{m+1}(z) + z B_{m+1}(z). \tag{4.17}$$

Given $A_{m+1}(z)$, the polynomial $B_{m+1}(z)$ is known, and so is

$$k_{m+1} = a^*_{m+1,m+1}. \tag{4.18}$$

Using this, we can uniquely identify $A_m(z)$, hence, $B_m(z)$, from (4.17). Thus, if we know the Nth-order optimal prediction polynomial $A_N(z)$, we can uniquely identify all the lower-order optimal predictors. So Eq. (4.17) is called the *downward* recursion. This process automatically reveals the lower-order lattice coefficients k_i. If we know the prediction error $\mathcal{E}^f_N$, we can therefore also identify the lower-order errors $\mathcal{E}^f_i$, by repeated use of Eq. (3.14), that is,

$$\mathcal{E}^f_m = \mathcal{E}^f_{m+1}/(1 - |k_{m+1}|^2) \tag{4.19}$$

4.4   CONCLUDING REMARKS

The properties of the IIR lattice have been analyzed in depth by Gray and Markel (1973), and lattice structures for linear prediction have been discussed in detail by Makhoul (1977). The advantage of lattice structures arises from the fact that even when the coefficients k_m are quantized the IIR lattice remains stable, as long as the quantized numbers continue to satisfy $|k_m| < 1$. This is a very useful result in compression and coding, because it guarantees stable reconstruction of the data segment that was compressed. We will return to this point in Sections 5.6 and 7.8.

· · · ·

CHAPTER 5

Autoregressive Modeling

5.1 INTRODUCTION

Linear predictive coding of a random process reveals a model for the process, called the autoregressive (AR) model. This model is very useful both conceptually and for approximating the process with a simple model. In this chapter, we describe this model.

5.2 AUTOREGRESSIVE PROCESSES

A WSS random process $w(n)$ is said to be autoregressive (AR) if it can be generated by using the *recursive difference equation*

$$w(n) = -\sum_{i=1}^{N} d_i^* w(n-i) + e(n), \qquad (5.1)$$

where

1. $e(n)$ is a zero-mean white WSS process, and
2. the polynomial $D(z) = 1 + \sum_{i=1}^{N} d_i^* z^{-i}$ has all zeros inside the unit circle.

In other words, we can generate $w(n)$ as the output of a causal, stable, all-pole IIR filter $1/D(z)$ in response to white input (Fig. 5.1). If $d_N \neq 0$, we say that the process is $AR(N)$, that is, AR of order N. Because $e(n)$ has zero mean, the AR process has zero mean according to the above definition.

Now, how does the AR process enter our discussion of linear prediction? Given a WSS process $x(n)$, let us assume that we have found the Nth-order optimal predictor polynomial $A_N(z)$. We know, we can then represent the process as the output of an IIR filter as shown in Fig. 2.1(b). The input to this filter is the prediction error $e_N^f(n)$. In the time domain, we can write

$$x(n) = -\sum_{i=1}^{N} a_{N,i}^* x(n-i) + e_N^f(n). \qquad (5.2)$$

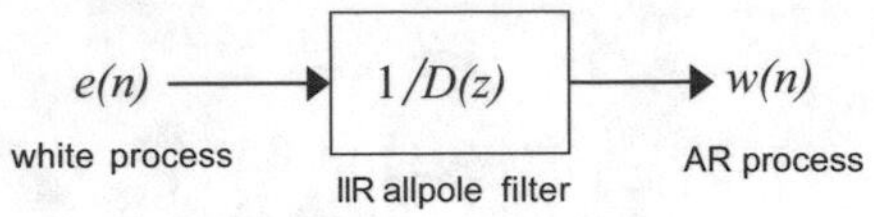

$e(n)$ → $1/D(z)$ → $w(n)$

white process   IIR allpole filter   AR process

FIGURE 5.1: Generation of an AR process from white noise and an all-pole IIR filter.

In Section 3.4, we saw that if the error stalls, that is, $\mathcal{E}_m^f$ does not decrease anymore as m increases beyond some value N, then $e_N^f(n)$ is white (assuming $x(n)$ has zero mean). Thus, the stalling phenomenon implies that $x(n)$ is $AR(N)$.

Summarizing, suppose the optimal predictors of various orders for a zero-mean process $x(n)$ are such that the minimized mean square errors satisfy

$$\mathcal{E}_1^f \geq \mathcal{E}_2^f \geq \ldots \geq \mathcal{E}_{N-1}^f > \mathcal{E}_N^f = \mathcal{E}_m^f, \quad m > N. \tag{5.3}$$

Then, $x(n)$ is $AR(N)$.

Example 5.1: Levinson's recursion for an AR process. Consider a WSS AR process generated by using an all-pole filter $c/D(z)$ with complex conjugate pairs of poles at

$$0.8e^{\pm 0.2 j\pi}, 0.85e^{\pm 0.4 j\pi}$$

so that

$$D(z) = 1.0000 - 1.8198z^{-1} + 2.0425z^{-2} - 1.2714z^{-3} + 0.4624z^{-4}.$$

The power spectrum $S_{ww}(e^{j\omega})$ of this AR(4) process is shown in Fig. 5.2 (top plot). The constant c in $c/D(z)$ is such that $S_{ww}(e^{j\omega})$ has maximum magnitude unity. The first few autocorrelation coefficients for this process are as follows:

$R(0)$	$R(1)$	$R(2)$	$R(3)$	$R(4)$	$R(5)$
0.2716	0.1758	−0.0077	−0.1091	−0.0848	−0.0226

If we perform Levinson's recursion with this, we obtain the lattice coefficients

k_1	k_2	k_3	k_4	k_5	k_6
−0.6473	0.7701	−0.5469	0.4624	0.0000	0.0000

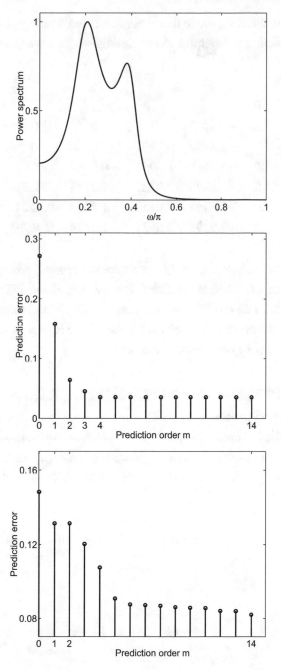

FIGURE 5.2: Example 5.1. The input power spectrum $S_{ww}(e^{j\omega})$ of an AR(4) process (top) and the prediction error $\mathcal{E}_m^f$ plotted for a few prediction orders m (middle). The prediction error for the non-AR process of Ex. 3.2 is also reproduced (bottom).

for the first six optimal predictors. Notice that these coefficients satisfy $|k_m| < 1$ as expected from theory. The optimal prediction polynomials $A_0(z)$ through $A_6(z)$ have the coefficients given in the following table:

$A_0(z)$	1.0						
$A_1(z)$	1.0	−0.6473	0	0	0	0	0
$A_2(z)$	1.0	−1.1457	0.7701	0	0	0	0
$A_3(z)$	1.0	−1.5669	1.3967	−0.5469	0	0	0
$A_4(z)$	1.0	−1.8198	2.0425	−1.2714	0.4624	0	0
$A_5(z)$	1.0	−1.8198	2.0425	−1.2714	0.4624	0	0
$A_6(z)$	1.0	−1.8198	2.0425	−1.2714	0.4624	0	0

The prediction filter does not change after $A_4(z)$ because the original process is AR(4). This is also consistent with the fact that the lattice coefficients past k_4 are all zero. The prediction errors are plotted in Fig. 5.2 (middle). The error $\mathcal{E}_m$ decreases monotonically and then becomes a constant for $m \geq 4$. For comparison, the prediction error $\mathcal{E}_m$ for the non-AR process of Ex. 3.2 is also shown in the figure (bottom); this error continues to decrease.

5.3   APPROXIMATION BY AN AR(N) PROCESSES

Recall that an arbitrary WSS process $x(n)$ (AR or otherwise), can always be represented in terms of the optimal linear prediction error $e_N^f(n)$ as in Eq. (5.2). In many practical cases, the error $e_N^f(n)$ tends to be nearly white for reasonably large N. If we replace $e_N^f(n)$ with zero-mean white noise $e(n)$ having the mean square value $\mathcal{E}_N^f$, we obtain the so-called AR model $y(n)$ for the process $x(n)$.

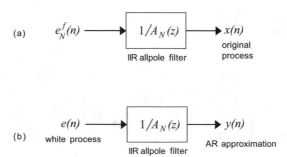

FIGURE 5.3: Generation of the $AR(N)$ approximation $y(n)$ for a WSS process $x(n)$. (a) The exact process $x(n)$ expressed in terms of the Nth-order LPC parameters and (b) the $AR(N)$ approximation $y(n)$.

This is shown in Fig. 5.3. The process $y(n)$ satisfies

$$y(n) = -\sum_{i=1}^{N} a_{N,i}^* y(n-i) + \underbrace{e(n)}_{\text{white}} \qquad (5.4)$$

The model process $y(n)$ is also called the $AR(N)$ approximation (Nth-order autoregressive approximation) of $x(n)$.

5.3.1 If a Process is AR, Then LPC Will Reveal It

If a zero-mean WSS process $x(n)$ is such that the optimal LPC error $e_N^f(n)$ is white, then we know that $x(n)$ is AR. Now, consider the converse. That is, let $x(n)$ be $AR(N)$. This means that we can express it as

$$x(n) = -\sum_{i=1}^{N} c_{N,i}^* x(n-i) + e(n), \qquad (5.5)$$

for some set of coefficients $c_{N,i}$, where $e(n)$ is white. This is a causal difference equation, so that, at any time n, the sample $x(n)$ is a linear combination

$$x(n) = g(0)e(n) + g(1)e(n-1) + g(2)e(n-2) + \dots \qquad (5.6)$$

Similarly, $x(n-i)$ is a linear combination of

$$e(n-i), e(n-i-1) \dots$$

Now consider

$$E[e(n)x^*(n-i)] = E\Big[e(n)[g(0)e(n-i) + g(1)e(n-i-1) + \dots]^*\Big]$$

Because $e(n)$ is white (i.e., $E[e(n)e^*(n-\ell)] = 0, \ell > 0$), this means that

$$E[e(n)x^*(n-i)] = 0, \quad i > 0. \qquad (5.7)$$

In other words, $e(n)$ is orthogonal to all the past samples of $x(n)$. In view of the orthogonality principle, this means that the linear combination

$$-\sum_{i=1}^{N} c_{N,i}^* x(n-i)$$

is the optimal linear prediction of $x(n)$. As the solution to the optimal Nth-order prediction problem is unique, the solution $\{a_{N,i}\}$ to the optimal predictor problem is therefore equal to $c_{N,i}$. Thus, the coefficients $c_{N,i}$ of the $AR(N)$ process can be identified simply by solving the Nth-order optimal prediction problem. The resulting prediction error $e_N^f(n)$ is white and equal to the "input term" $e(n)$ in the AR equation Eq. (5.5). Summarizing, we have:

Theorem 5.1. *LPC of an AR process.* Let $x(n)$ be an $AR(N)$ WSS process, that is, a zero-mean process representable as in Eq. (5.5), where $e(n)$ is white. Then,

1. The AR coefficients $c_{N,i}$ can be found simply by performing Nth-order LPC on the process $x(n)$. The optimal LPC coefficients $a_{N,i}$ will turn out to be $c_{N,i}$. Thus, the AR model $y(n)$ resulting from the optimal LPC (Fig. 5.3) is the same as $x(n)$, that is $y(n) = x(n)$.
2. Moreover, the prediction error $e_N^f(n)$ is white and the same as the white noise sequence $e(n)$ appearing in the $AR(N)$ description Eq. (5.5) of $x(n)$. $\diamond$

5.3.2   Extrapolation of the Autocorrelation

Because $e_N^f(n)$ is white, the autocorrelation $R(k)$ of an $AR(N)$ process $x(n)$ is completely determined by the filter $1/A_N(z)$ and the variance $\mathcal{E}_N^f$. Because the quantities $A_N(z)$ and $\mathcal{E}_N^f$ can be found from the $N + 1$ coefficients

$$R(0), R(1), \ldots, R(N)$$

using Levinson's recursion, we conclude that for $AR(N)$ processes, all the coefficients $R(k)$, $|k| > N$ are determined by the above $N + 1$ coefficients. So, the determination of $A_N(z)$ and $\mathcal{E}_N^f$ provides us a way to extrapolate an autocorrelation. In other words, we can find $R(k)$ for $|k| > N$ from the first $N + 1$ values in such a way that *the Fourier transform $S_{xx}(e^{j\omega})$ of the extrapolated sequence is nonnegative for all ω.*

Extrapolation equation. An explicit equation for extrapolation of $R(k)$ can readily be written down. Thus, given $R(0), R(1), \ldots R(N)$ for an $AR(N)$ process, we can find $R(k)$ for any $k > N$ using the equation

$$R(k) = -a_{N,1}^* R(k-1) - a_{N,2}^* R(k-2) \ldots - a_{N,N}^* R(k-N).$$

This equation follows directly from normal equations. See Appendix D for details. $\square$

In many practical situations, a signal $x(n)$ that is not AR can be modeled well by an AR process. A common example is speech (Markel and Gray, 1976; Rabiner and Schafer, 1978; Jayant and Noll, 1984). The fact that $x(n)$ is not AR means that the prediction error $e_m^f(n)$ never gets white, but its power spectrum often gets flatter and flatter as m increases. This will be demonstrated later in Section 6.3 where we define a mathematical measure for spectral flatness and study it in the context of linear prediction.

5.4 AUTOCORRELATION MATCHING PROPERTY

Let $x(n)$ be a WSS process, $A_N(z)$ be its optimal Nth-order predictor polynomial, and $y(n)$ be the $AR(N)$ approximation of $x(n)$ generated as in Fig. 5.3. In what mathematical sense does $y(n)$ approximate $x(n)$, that is, in what respect is $y(n)$ "close" to $x(n)$? We now provide a quantitative answer:

Theorem 5.2. Let $R(k)$ and $r(k)$ be the autocorrelations of $x(n)$ and the $AR(N)$ approximation $y(n)$, respectively. Then,

$$R(k) = r(k), \quad |k| \leq N. \tag{5.8}$$

Thus, the $AR(N)$ model $y(n)$ is an approximation of $x(n)$ in the sense that the first $N+1$ autocorrelation coefficients for the two processes are equal to each other. ◇

So, as the approximation order N increases, more and more values of $R(k)$ are matched to $r(k)$. If the process $x(n)$ is itself $AR(N)$, then we know (Theorem 5.1) that the $AR(N)$ model $y(n)$ satisfies $y(n) = x(n)$ so that Eq. (5.8) holds for all k, not just $|k| \leq N$. A simple corollary of Theorem 5.2 is

$$R(0) = r(0),$$

which can be used to prove that for any WSS process $x(n)$,

$$R(0) = \mathcal{E}_N^f \int_0^{2\pi} \frac{1}{|A_N(e^{j\omega})|^2} \frac{d\omega}{2\pi} \tag{5.9}$$

To see this, observe first that the AR process $y(n)$ in Fig. 5.3(b) is generated using white noise $e(n)$ with variance $\mathcal{E}_N^f$, so that

$$r(0) = \int_0^{2\pi} \frac{\mathcal{E}_N^f}{|A_N(e^{j\omega})|^2} \frac{d\omega}{2\pi}.$$

But because $r(0) = R(0)$, Eq. (5.9) follows readily.

Proof of Theorem 5.2. By construction, the $AR(N)$ process $y(n)$ is representable as in Fig. 5.3(b) and therefore satisfies the difference equation Eq. (5.4). Now consider the sample $y(n-k)$. Because Fig. 5.3(b) is a causal system,

$$y(n-k) = \text{ linear combination of } e(n-k-i), \quad i \geq 0. \tag{5.10}$$

But because $e(n)$ is zero-mean white, we have $E[e(n)e^*(n-k)] = 0$ for $k \neq 0$. Thus,

$$E[e(n)y^*(n-k)] = 0, \quad k > 0. \tag{5.11}$$

Multiplying Eq. (5.4) by $y^*(n-k)$, taking expectations, and using the above equality, we obtain the following set of equations.

$$
\begin{bmatrix}
r(0) & r(1) & \cdots & r(N) \\
r^*(1) & r(0) & \cdots & r(N-1) \\
\vdots & \vdots & \ddots & \vdots \\
r^*(N) & r^*(N-1) & \cdots & r(0)
\end{bmatrix}
\begin{bmatrix}
1 \\
a_{N,1} \\
\vdots \\
a_{N,N}
\end{bmatrix}
=
\begin{bmatrix}
\mathcal{E}_N^f \\
0 \\
\vdots \\
0
\end{bmatrix}
\tag{5.12}
$$

where $r(k) = E[y(n)y^*(n-k)]$. This set of equations has exactly the same appearance as the augmented normal equations Eq. (2.20). Thus, the coefficients $a_{N,i}$ are the solutions to two optimal predictor problems: one based on the process $x(n)$ and the other based on $y(n)$.

We now claim that Eq. (2.20) and (5.12) together imply $R(k) = r(k)$, for the range of values $0 \le k \le N$. This is proved by observing that the coefficients $a_{N,i}$ and $\mathcal{E}_N^f$ completely determine the lower-order optimal prediction coefficients $a_{m,i}$ and errors $\mathcal{E}_m^f$ (see end of Section 4.3.3, "Upward and Downward Recursions"). Thus, $a_{m,i}$ and $\mathcal{E}_m^f$ satisfy smaller sets of equations of the form Eq. (2.20). Collecting them together, we obtain a set of equations resembling Eq. (2.37). We obtain a similar relation if $R(k)$ is replaced by $r(k)$. Thus, we have the following two sets of equations (demonstrated for $N=3$)

$$
\underbrace{
\begin{bmatrix}
r(0) & r(1) & r(2) & r(3) \\
r^*(1) & r(0) & r(1) & r(2) \\
r^*(2) & r^*(1) & r(0) & r(1) \\
r^*(3) & r^*(2) & r^*(1) & r(0)
\end{bmatrix}
}_{\mathbf{r}_{N+1}}
\underbrace{
\begin{bmatrix}
1 & 0 & 0 & 0 \\
a_{3,1} & 1 & 0 & 0 \\
a_{3,2} & a_{2,1} & 1 & 0 \\
a_{3,3} & a_{2,2} & a_{1,1} & 1
\end{bmatrix}
}_{\boldsymbol{\Delta}_\ell}
=
\underbrace{
\begin{bmatrix}
\mathcal{E}_3^f & \times & \times & \times \\
0 & \mathcal{E}_2^f & \times & \times \\
0 & 0 & \mathcal{E}_1^f & \times \\
0 & 0 & 0 & \mathcal{E}_0^f
\end{bmatrix}
}_{\boldsymbol{\Delta}_{u_1}}
$$

$$
\underbrace{
\begin{bmatrix}
R(0) & R(1) & R(2) & R(3) \\
R^*(1) & R(0) & R(1) & R(2) \\
R^*(2) & R^*(1) & R(0) & R(1) \\
R^*(3) & R^*(2) & R^*(1) & R(0)
\end{bmatrix}
}_{\mathbf{R}_{N+1}}
\underbrace{
\begin{bmatrix}
1 & 0 & 0 & 0 \\
a_{3,1} & 1 & 0 & 0 \\
a_{3,2} & a_{2,1} & 1 & 0 \\
a_{3,3} & a_{2,2} & a_{1,1} & 1
\end{bmatrix}
}_{\boldsymbol{\Delta}_\ell}
=
\underbrace{
\begin{bmatrix}
\mathcal{E}_3^f & \times & \times & \times \\
0 & \mathcal{E}_2^f & \times & \times \\
0 & 0 & \mathcal{E}_1^f & \times \\
0 & 0 & 0 & \mathcal{E}_0^f
\end{bmatrix}
}_{\boldsymbol{\Delta}_{u_2}}
$$

The symbol $\times$ stands for possibly nonzero entries whose values are irrelevant for the discussion. Note that $\boldsymbol{\Delta}_\ell$ is lower triangular and $\boldsymbol{\Delta}_{u_1}$ and $\boldsymbol{\Delta}_{u_2}$ are upper triangular matrices. We have not shown that $\Delta_{u_1} = \Delta_{u_2}$, so we will take a different route to prove the desired claim Eq. (5.8). By premultiplying the preceding equations with $\Delta_\ell^\dagger$, we get

$$\Delta_\ell^\dagger \mathbf{r}_{N+1} \Delta_\ell = \Delta_\ell^\dagger \Delta_{u_1},$$
$$\Delta_\ell^\dagger \mathbf{R}_{N+1} \Delta_\ell = \Delta_\ell^\dagger \Delta_{u_2}.$$

The right-hand side in each of these equations is a product of two upper triangular matrices and is therefore upper triangular. The left-hand sides, on the other hand, are Hermitian. As a result, the right-hand sides *must* be diagonal! It is easy to verify that these diagonal matrices have $\mathcal{E}_m^f$ as diagonal elements, so that the right-hand sides of the above two equations are identical. This proves that $\Delta_\ell^\dagger \mathbf{R}_{N+1} \Delta_\ell = \Delta_\ell^\dagger \mathbf{r}_{N+1} \Delta_\ell$. Since Δ_ℓ is nonsingular, we can invert it, and obtain $\mathbf{R}_{N+1} = \mathbf{r}_{N+1}$. This proves Eq. (5.8) indeed. $\qquad\square$

In a nutshell, given $\{a_{N,i}\}$ and the error $\mathcal{E}_N^f$, we can use the downward recursion (Section 4.3.3) to compute $a_{m,i}$ and $\mathcal{E}_m^f$ for $m = N - 1, N - 2, \ldots$ So we uniquely identify $R(0) = \mathcal{E}_0^f$. Next, from the normal equation with $m = 1$, we can identify $R(1)$. In general, if $R(0), \ldots, R(m-1)$ are known, we can use the normal equation for mth order and identify $R(m)$ from the mth equation in this set (refer to Eq. (2.8)). Thus, the coefficients $R(0), R(1), \ldots R(N)$, which lead to a given set of $\{a_{N,i}\}$ and $\mathcal{E}_N^f$, are *unique*, so $r(k)$ and $R(k)$ in Eqs. (2.20) and (5.12) must be the same!

R(k) is Matched to the Autocorrelation of the IIR Filter. The $AR(N)$ model of Fig. 5.3(b) can be redrawn as in Fig. 5.4 where $e_1(n)$ is white noise with variance equal to unity and where

$$H(z) = \frac{\sqrt{\mathcal{E}_N^f}}{A_N(z)} \qquad (5.13)$$

$H(z)$ is a causal IIR filter with impulse response, say, $h(n)$. The deterministic autocorrelation of the sequence $h(n)$ is defined as

$$\sum_n h(n) h^*(n-k).$$

Because $e_1(n)$ is white with $E[|e_1(n)|^2] = 1$, the autocorrelation $r(k)$ of the AR model process $y(n)$ is given by

$$r(k) = \sum_n h(n) h^*(n-k). \qquad (5.14)$$

So we can rephrase Eq. (5.8) as follows:

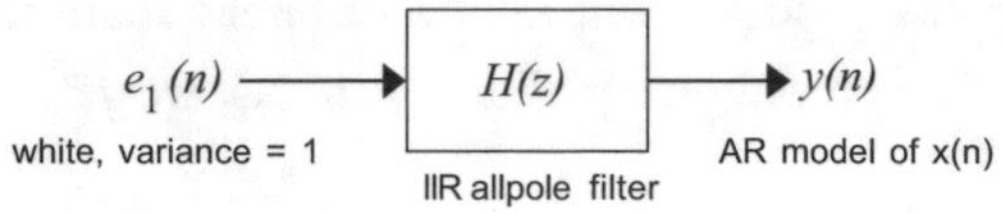

$$e_1(n) \longrightarrow \boxed{H(z)} \longrightarrow y(n)$$

white, variance = 1   AR model of x(n)

IIR allpole filter

FIGURE 5.4: Autoregressive model $y(n)$ of a signal $x(n)$.

Corollary 5.1. Let $x(n)$ be a WSS process with autocorrelation $R(k)$ and $A_N(z)$ be the Nth-order optimal predictor polynomial. Let $h(n)$ be the impulse response of the causal stable IIR filter $\sqrt{\mathcal{E}_N^f}/A_N(z)$. Then,

$$R(k) = \sum_{n=0}^{\infty} h(n) h^*(n-k), \qquad (5.15)$$

for $|k| \leq N$. That is, the first $N+1$ coefficients of the deterministic autocorrelation of $h(n)$ are matched to the values $R(k)$. In particular, if $x(n)$ is $AR(N)$, then Eq. (5.15) holds for all k.   ◇

5.5   POWER SPECTRUM OF THE AR MODEL

We just proved that the $AR(N)$ approximation $y(n)$ is such that its autocorrelation coefficients $r(k)$ match that of the original process $x(n)$ for the first $N+1$ values of k. As N increases, we therefore expect the power spectrum $S_{yy}(e^{j\omega})$ of $y(n)$ to resemble the power spectrum $S_{xx}(e^{j\omega})$ of $x(n)$. This is demonstrated next.

Example 5.2: AR approximations. In this example, we consider a random process generated as the output of the fourth-order IIR filter

$$G_4(z) = \frac{3.9 - 2.7645z^{-1} + 1.4150z^{-2} - 0.5515z^{-3}}{1 - 0.9618z^{-1} + 0.7300z^{-2} - 0.5315z^{-3} + 0.5184z^{-4}}$$

This system has complex conjugate pairs of poles inside the unit circle at $0.9e^{\pm 0.2\,j\pi}$ and $0.8e^{\pm 0.6\,j\pi}$. With white noise of variance σ_e^2 driving this system, the output has the power spectrum

$$S_{xx}(e^{j\omega}) = \sigma_e^2 |G_4(e^{j\omega})|^2.$$

This is shown in Fig. 5.5 (dotted plots), with σ_e^2 assumed to be such that the peak value of $S_{xx}(e^{j\omega})$ is unity. Clearly $S_{xx}(e^{j\omega})$ is not an AR process because the numerator of $G_4(z)$ is not a constant.

Figure 5.5 shows the $AR(N)$ approximations $S_{yy}(e^{j\omega})$ of the spectrum $S_{xx}(e^{j\omega})$ for various orders N (solid plots). We see that the $AR(4)$ approximation is quite unacceptable, whereas the $AR(5)$ approximation shows significant improvement. The successive approximations are better and better until we find that the $AR(9)$ approximation is nearly perfect. From the preceding theory, we know that the $AR(9)$ approximation and the original process have matching

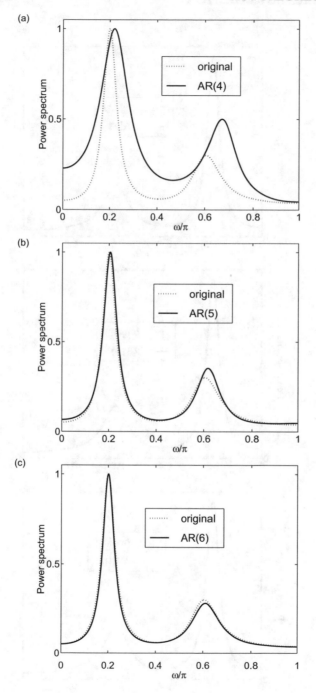

FIGURE 5.5: AR approximations of a power spectrum: (a) AR(4), (b) AR(5), (c) AR(6), (d) AR(7), (e)AR(8), and (f) AR(9).

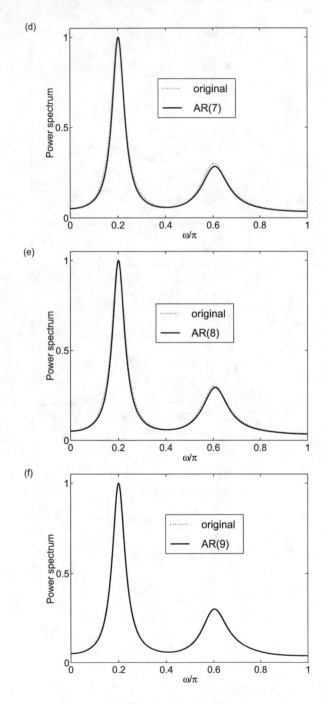

FIGURE 5.5: Continued.

autocorrelation coefficients, that is, $R(m) = r(m)$ for $0 \leq m \leq 9$. This is indeed seen from the following table:

m	$R(m)$	$r(m)$
0	0.1667	0.1667
1	0.0518	0.0518
2	−0.0054	−0.0054
3	0.0031	0.0031
4	−0.0519	−0.0519
5	−0.0819	−0.0819
6	−0.0364	−0.0364
7	−0.0045	−0.0045
8	0.0057	0.0057
9	0.0318	0.0318
10	0.0430	0.0441
11	0.0234	0.0241

The coefficients $R(m)$ and $r(m)$ begin to differ starting from $m = 10$ as expected. So the $AR(9)$ approximation is not perfect either, although the plots of $S_{xx}(e^{j\omega})$ and $S_{yy}(e^{j\omega})$ are almost indistinguishable.

The spectrum $S_{yy}(e^{j\omega})$ of the AR approximation $y(n)$ is called an *AR-model-based estimate* of $S_{xx}(e^{j\omega})$. This is also said to be the *maximum entropy estimate*, for reasons explained later in Section 6.6.1. We can say that the estimate $S_{yy}(e^{j\omega})$ is obtained by extrapolating the finite autocorrelation segment

$$R(k), \quad |k| \leq N \tag{5.16}$$

using an $AR(N)$ model. The estimate $S_{yy}(e^{j\omega})$ is nothing but the Fourier transform of the extrapolated autocorrelation $r(k)$. Note that if $R(k)$ were "extrapolated" by forcing it to be zero for $|k| > N$, then the result may not even be a valid autocorrelation (i.e., its Fourier transform may not be nonnegative everywhere). Further detailed discussions on power spectrum estimation techniques can be found in Kay and Marple (1981), Marple (1987), Kay (1988), and Therrien (1992).

Peaks are well matched. We know that the $AR(N)$ model approximates the power spectrum $S_{xx}(e^{j\omega})$ with the all-pole spectrum power spectrum

$$S_{yy}(e^{j\omega}) = \frac{\mathcal{E}_N^f}{|A_N(e^{j\omega})|^2} \tag{5.17}$$

If $S_{xx}(e^{j\omega})$ has sharp peaks, then $A_N(z)$ has to have zeros close to the unit circle to approximate these peaks. If, on the other hand, $S_{xx}(e^{j\omega})$ has zeros (or sharp dips) on the unit circle, then $S_{yy}(e^{j\omega})$ cannot approximate these very well because it cannot have zeros on the unit circle. Thus, the $AR(N)$ model can be used to obtain a good match of the power spectrum

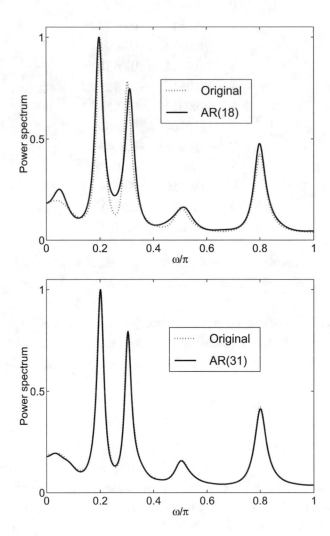

FIGURE 5.6: Demonstrating that the AR approximation of a power spectrum shows good agreement near the peaks. In the AR(18) approximation, the peaks are more nicely matched than the valleys. The AR(31) approximation is nearly perfect. The original power spectrum is generated with a pole-zero filter as in Ex. 5.2, but instead of $G_4(z)$, we have in this example a 10th-order real-coefficient filter $G_{10}(z)$ with all poles inside the unit circle.

$S_{xx}(e^{j\omega})$ near its peaks but not near its valleys. This is demonstrated in the example shown in Fig. 5.6. Such approximations, however, are useful in some applications such as the AR representation of speech. □

Spectral factorization. If the process $x(n)$ is indeed $AR(N)$, then the autoregressive model is such that $S_{xx}(e^{j\omega}) = S_{yy}(e^{j\omega})$. So,

$$S_{xx}(e^{j\omega}) = \frac{\mathcal{E}_N^f}{|A_N(e^{j\omega})|^2} \tag{5.18}$$

In other words, we have found a rational transfer function $H_N(z) = c/A_N(z)$ such that

$$S_{xx}(e^{j\omega}) = |H_N(e^{j\omega})|^2.$$

We see that $H_N(z)$ is a spectral factor of $S_{xx}(e^{j\omega})$. Thus, when $x(n)$ is $AR(N)$, the Nth-order prediction problem places in evidence a stable spectral factor $H_N(z)$ of the power spectrum $S_{xx}(e^{j\omega})$. When $x(n)$ is not $AR(N)$, the filter $H_N(z)$ is an $AR(N)$ approximation to the spectral factor.

Example 5.3: Estimation of AR model paramaters. We now consider an $AR(4)$ process $x(n)$ derived by driving the filter $1/A_4(z)$ with white noise, where

$$A_4(z) = 1.0000 - 1.8198z^{-1} + 2.0425z^{-2} - 1.2714z^{-3} + 0.4624z^{-4}$$

This has roots

$$0.2627 \pm 0.8084\,j, \ 0.6472 \pm 0.4702\,j$$

with absolute values 0.85 and 0.8. A sample of the $AR(N)$ process can be generated by first generating a WSS white process $e(n)$ using the Matlab command *randn* and driving the filter $1/A_4(z)$ with input $e(n)$. The purpose of the example is to show that, from the measured values of the random process $x(n)$, we can actually estimate the autocorrelation and hence, the model parameters (coefficients of $A_4(z)$) accurately.

First, we estimate the autocorrelation $R(k)$ using (2.43), where L is the number of samples of $x(n)$ available. To see the effect of noise (which is always present in practice), we add random noise to the output $x(n)$, so the actual data used are

$$x_{\text{noisy}}(n) = x(n) + \eta(n).$$

The measurement noise $\eta(n)$ is assumed to be zero-mean white Gaussian. Its variance will be denoted as σ_η^2. With $L = 2^{10}$ and $\sigma_\eta^2 = 10^{-4}$, and $x_{\text{noisy}}(n)$ used instead of $x(n)$, the estimated autocorrelation $R(k)$ for $0 \leq k \leq 4$ is as follows:

$$R(0) = 8.203, R(1) = 4.984, R(2) = -1.032, R(3) = -3.878, R(4) = -2.250.$$

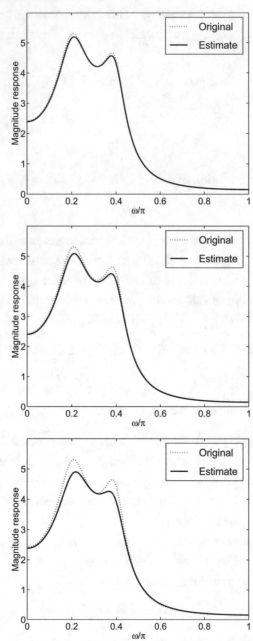

FIGURE 5.7: Example 5.3. Estimation of AR model parameters. Magnitude responses of the original system $1/A_4(z)$ and the estimate $1/\widehat{A}_4(z)$ are shown for three values of the measurement noise variance $\sigma_\eta^2 = 10^{-4}$, 0.0025, and $\sigma_\eta^2 = 0.01$, respectively, from top to bottom. The segment length used to estimate autocorrelation is $L = 2^{10}$.

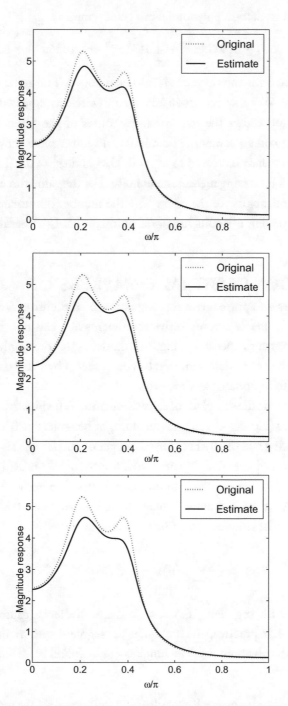

FIGURE 5.8: Example 5.3. Repetition of Fig. 5.7 with segment length $L = 2^8$ instead of 2^{10}.

From this, the estimated prediction polynomial can be obtained as

$$\widehat{A}_4(z) = 1.0000 - 1.8038z^{-1} + 2.0131z^{-2} - 1.2436z^{-3} + 0.4491z^{-4}$$

This should be compared with the original $A_4(z)$ given above. The roots of $\widehat{A}_4(z)$ are $0.2616 \pm 0.8036\,j$ and $0.6403 \pm 0.4679\,j$, which are inside the unit circle as expected.

Figure 5.7 (top plot) shows the magnitude responses of the filters $1/A_4(z)$ and $1/\widehat{A}_4(z)$, which are seen to be in good agreement. The middle and lower plots show these responses with the measurement noise variance increased to $\sigma_\eta^2 = 0.0025$ and $\sigma_\eta^2 = 0.01$, respectively. Thus, the estimate deteriorates with increasing measurement noise. For the same three levels of measurement noise, Fig. 5.8 shows similar plots for the case where the number of measured samples of $x(n)$ are reduced to $L = 2^8$. Decreasing L results in poorer estimates of $A_4(z)$ because the estimated $R(k)$ is less accurate.

5.6   APPLICATION IN SIGNAL COMPRESSION

The $AR(N)$ model gives an approximate representation of the entire WSS process $x(n)$, using only $N + 1$ numbers, (i.e., the N autoregressive coefficients $a_{N,i}^*$ and the mean square value $\mathcal{E}_N^f$). In practical applications such as speech coding, this modeling is commonly used for compressing the signal waveform. Signals of practical interest can be modeled by WSS processes only over short periods. The model has to be updated as time evolves.

It is typical to subdivide the sampled speech waveform $x(n)$ (evidently, a real-valued sequence) into consecutive segments (Fig. 5.9).[1] Each segment might be typically 20-ms long, corresponding to 200 samples at a sampling rate of 10 kHz. This sequence of 200 samples is viewed as representing a WSS random process, and the first N autocorrelation coefficients $R(k)$ are estimated. With $N << 200$ (usually $N \approx 10$), the finiteness of duration of the segment is of secondary importance in affecting the accuracy of the estimates. There are several sophisticated ways to perform this estimation (Section 2.5). The simplest would be to use the average

$$R(k) \approx \frac{1}{L} \sum_{n=0}^{L-1} x(n)x(n-k), \quad 0 \le k \le N. \tag{5.19}$$

Here, L is the length of the segment. If $N << L$, the end effects (caused by the use of finite summation) are negligible. In practice, one multiplies the segment with a smooth window to reduce the end effects. Further details about these techniques can be found in Rabiner and Schafer (1978).

---

[1]Note that when $x(n)$ is real-valued as in speech, the polynomial coefficients $a_{N,i}$, the parcor coefficients k_i, and the prediction error sequence $e_N^f(n)$ are real as well.

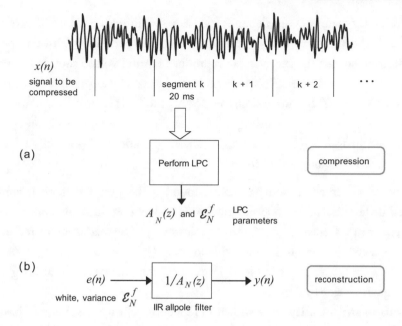

FIGURE 5.9: (a) Compression of a signal segment using LPC and (b) reconstruction of an approxima-tion of the original segment from the LPC parameters $A_N(z)$ and $\mathcal{E}_N^f$. The kth segment of the original signal $x(n)$ is approximated by a segment of the AR signal $y(n)$. For the next segment of $x(n)$, the LPC parameters $A_N(z)$ and $\mathcal{E}_N$ are generated again.

Note that with segment length $L = 200$ and the AR model order $N = 10$, a segment of 200 samples has been compressed into 11 coefficients! The encoded message

$$\left[a_{N,1} \; a_{N,2} \; \cdots \; a_{N,N} \; \mathcal{E}_N^f \right] \qquad (5.20)$$

is then transmitted, possibly after quantization. At the receiver end, one generates a local white-noise source $e(n)$ with variance $\mathcal{E}_N^f$ and drives the IIR filter

$$\frac{1}{A_N(z)} = \frac{1}{1 + \sum_{i=1}^{N} a_{N,i} z^{-i}} \qquad (5.21)$$

with this noise source as in Fig. 5.9(b). A segment of the output $y(n)$ is then taken as the $AR(N)$ approximation of the original speech segment.

Pitch-excited coders and noise-excited coders. Strictly speaking, the preceding discussion tells only half the story. The locally generated white noise $e(n)$ is satisfactory only when the speech segment under consideration is *unvoiced*. Such sounds have no pitch or periodicity (e.g., the part

"Sh" in "Should I." By contrast, *voiced* sounds such as vowel utterances (e.g., "I") are characterized by a pitch, which is the local periodicity of the utterance. This periodicity T of the speech segment has to be transmitted so that the receiver can reproduce faithful voiced sounds. The modification required at the receiver is that the $AR(N)$ model is now driven by a periodic waveform such as an impulse train, with period T. Summarizing, the $AR(N)$ model is excited by a noise generator or a periodic impulse train generator, according to whether the speech segment is unvoiced or voiced. This is summarized in Fig. 5.10. There are further complications in practice such as transitions from voiced to unvoiced regions and so forth.   □

Many variations of linear prediction have been found to be useful in speech compression. A technique called differential pulse code modulation is widely used in many applications; this uses the predictive coder in a feedback loop in an ingenious way. The reader interested in the classical literature on speech compression techniques should study (Atal and Schroeder, 1970; Rabiner and Schafer, 1978; Jayant and Noll, 1984; Deller et al., 1993) or the excellent, although old, tutorial paper on vector quantization by Makhoul et al. (1985).

      Quantization and stability. It is shown in Appendix C that the unquantized coefficients $a_{N,i}$, which result from the normal equations are such that $1/A_N(z)$ has all poles inside the unit circle (i.e., it is causal and stable). However, if we quantize $a_{N,i}$ before transmission, this may not continue to remain true, particularly in low bit-rate coding. A solution to this problem is to transmit the "parcor" or lattice coefficients k_m. From Section 3.3, we know that $|k_m| < 1$. In view of the relation between Levinson's recursion and lattice structures (Section 4.3.1), we also know that $1/A_N(z)$ is stable if and only if $|k_m| < 1$. Suppose the coefficients k_m are quantized in such a way that $|k_m| < 1$ continues to hold after quantization. At the receiver end, one recomputes an approximation $A_N^{(q)}(z)$ of $A_N(z)$, from the quantized versions k_m. Then, the $AR(N)$ model $1/A_N^{(q)}(z)$ is guaranteed to be stable.

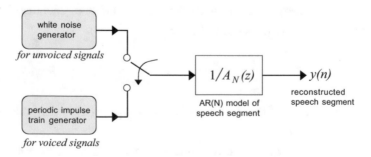

FIGURE 5.10: Reconstruction of a speech segment from its $AR(N)$ model. The source of excitation depends on whether the segment is voiced or unvoiced (see text).

In Section 7.8, we introduce the idea of *line-spectrum pair* (LSP). Instead of quantizing and transmitting the coefficients of $A_N(z)$ or the lattice coefficients k_m, one can quantize and transmit the LSPs. Like the lattice coefficients, the LSP coefficients preserve stability even after quantization. Other advantages of using the LSP coefficients will be discussed in Section 7.8.

5.7   MA AND ARMA PROCESSES

We know that a WSS random process $x(n)$ is said to be AR if it satisfies a recursive (IIR) difference equation of the form

$$x(n) = -\sum_{i=1}^{N} d_i^* x(n-i) + e(n), \tag{5.22}$$

where $e(n)$ is a zero-mean white WSS process, and the polynomial $D(z) = 1 + \sum_{i=1}^{N} d_i^* z^{-i}$ has all zeros inside the unit circle. We say that a WSS process $x(n)$ is a moving average (MA) process if it satisfies a nonrecursive (FIR) difference equation of the form

$$x(n) = \sum_{i=0}^{N} p_i^* e(n-i), \tag{5.23}$$

where $e(n)$ is a zero-mean white WSS process. Finally, we say that a WSS process $x(n)$ is an ARMA process if

$$x(n) = -\sum_{i=1}^{N} d_i^* x(n-i) + \sum_{i=0}^{N} p_i^* e(n-i), \tag{5.24}$$

where $e(n)$ is a zero-mean white WSS process. Defining the polynomials

$$D(z) = 1 + \sum_{i=1}^{N} d_i^* z^{-i} \quad \text{and} \quad P(z) = \sum_{i=0}^{N} p_i^* z^{-i}, \tag{5.25}$$

we see that the above processes can be represented as in Fig. 5.11. In each of the three cases, $x(n)$ is the output of a rational discrete time filter, driven by zero-mean white noise. For the AR process, the filter is an all-pole filter. For the MA process, the filter is FIR. For the ARMA process, the filter is IIR with both poles and zeros.

We will not have occassion to discuss MA and ARMA processes further. However, the theory and practice of linear prediction has been modified to obtain approximate models for these types of processes. Some discussions can be found in Marple (1987) and Kay (1988). In Section 5.5, we saw that the AR model provides an exact spectral factorization for an AR process. An

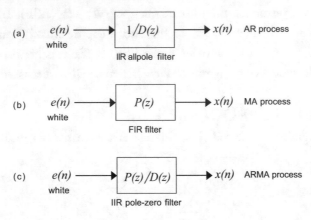

FIGURE 5.11: (a) AR, (b) MA, and (c) ARMA processes.

approximate spectral factorization algorithm for MA processes, based on the theory of linear prediction, can be found in Friedlander (1983).

5.8   SUMMARY

We conclude the chapter with a summary of the properties of AR processes.

1. An AR process $x(n)$ of order N ($AR(N)$ process) satisfies an Nth-order recursive difference equation of the form Eq. (5.22), where $e(n)$ is zero-mean white noise. Equivalently, $x(n)$ is the output of a causal stable IIR all-pole filter $1/D(z)$ (Fig. 5.11(a)) excited by white noise $e(n)$.

2. If we perform Nth-order linear prediction on a WSS process $x(n)$, we obtain the representation of Fig. 5.3(a), where $e_N^f(n)$ is the optimal prediction error and $A_N(z)$ is the predictor polynomial. By replacing $e_N^f(n)$ with white noise, we obtain the $AR(N)$ model of Fig. 5.3(b). Here, $y(n)$ is an $AR(N)$ process, and is the $AR(N)$ approximation (or model) for $x(n)$.

3. The $AR(N)$ model $y(n)$ approximates $x(n)$ in the sense that the autocorrelations of the two processes, denoted $r(k)$ and $R(k)$, respectively, are matched for $0 \leq |k| \leq N$ (Theorem 5.2).

4. The $AR(N)$ model $y(n)$ is such that its power spectrum $S_{yy}(e^{j\omega})$ tends to approximate the power spectrum $S_{xx}(e^{j\omega})$ of $x(n)$. This approximation improves as N increases and is particularly good near the peaks of $S_{xx}(e^{j\omega})$ (Fig. 5.6). Near the valleys of $S_{xx}(e^{j\omega})$, it is not so good, because the AR spectrum $S_{yy}(e^{j\omega})$ is an all-pole spectrum and cannot have zeros on or near the unit circle.

5. If $x(n)$ itself is $AR(N)$, then the optimal Nth-order prediction error $e_N^f(n)$ is white. Thus, $x(n)$ can be written as in Eq. (5.22), where $e(n)$ is white and d_i are the solutions (usually denoted $a_{N,i}$) to the normal equations Eq. (2.8). Furthermore, the $AR(N)$ power spectrum $\mathcal{E}_N^f / |A_N(e^{j\omega})|^2$ is exactly equal to $S_{xx}(e^{j\omega})$. Thus, we can compute the power spectrum $S_{xx}(e^{j\omega})$ exactly, simply by knowing the autocorrelation coefficients $R(k)$, $|k| \leq N$ and identifying $A_N(z)$ using normal equations. This means, in turn, that all the autocorrelation coefficients $R(k)$ of an $AR(N)$ process $x(n)$ are determined, if $R(k)$ is known for $|k| \leq N$.

·   ·   ·   ·   ·

CHAPTER 6

Prediction Error Bound and Spectral Flatness

6.1 INTRODUCTION

In this chapter, we first derive a closed-form expression for the minimum mean-squared prediction error for an $AR(N)$ process. We then introduce an important concept called spectral flatness. This measures the extent to which the power spectrum $S_{xx}(e^{j\omega})$ of a WSS process $x(n)$, not necessarily AR, is "flat." The flatness γ_x^2 is such that $0 < \gamma_x^2 \leq 1$, with $\gamma_x^2 = 1$ for a flat power spectrum (white process). We will then prove two results:

1. For fixed mean square value $R(0)$, as the flatness measure of an $AR(N)$ process $x(n)$ gets smaller, the mean square optimal prediction error $\mathcal{E}_N^f$ also gets smaller (Section 6.4). So, a *less flat* process is *more predictable*.
2. For any WSS process (not necessarily AR), the flatness measure of the optimal prediction error $e_m^f(n)$ *increases* as m increases. So, the power spectrum of the prediction error gets flatter and flatter as the prediction order m increases (Section 6.5).

6.2 PREDICTION ERROR FOR AN AR PROCESS

If $x(n)$ is an $AR(N)$ process, we know that the optimal prediction error $e_N^f(n)$ is white. We will show that the mean square error $\mathcal{E}_N^f$ can be expressed in closed form as

$$\mathcal{E}_N^f = \exp\left(\frac{1}{2\pi} \int_0^{2\pi} \text{Ln} S_{xx}(e^{j\omega}) \, d\omega\right) \qquad (6.1)$$

where $S_{xx}(e^{j\omega})$ is the power spectrum of $x(n)$. Here, Ln is the natural logarithm of a positive number as normally defined in elementary calculus and demonstrated in Fig. 6.1. Because $x(n)$ is AR, the spectrum $S_{xx}(e^{j\omega}) > 0$ for all ω, and $\text{Ln} S_{xx}(e^{j\omega})$ is real and finite for all ω. The derivation of the above expression depends on the following lemma.

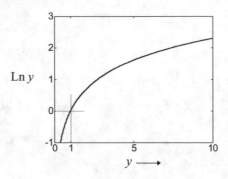

FIGURE 6.1: A plot of the logarithm.

Lemma 6.1. Consider a polynomial in z^{-1} of the form $A(z) = 1 + \sum_{n=1}^{N} a_n z^{-n}$, and let all the zeros of $A(z)$ be strictly inside the unit circle. Then,

$$\frac{1}{2\pi} \int_0^{2\pi} \ln A(e^{j\omega}) \, d\omega = j2\pi k \tag{6.2}$$

for some integer k. This implies $\int_0^{2\pi} \ln|A(e^{j\omega})|^2 d\omega / 2\pi = j2\pi m$ for some integer m, so that

$$\frac{1}{2\pi} \int_0^{2\pi} \text{Ln}|A(e^{j\omega})|^2 d\omega = 0, \tag{6.3}$$

where $\text{Ln}(.)$ stands for the principal value of the logarithm.   ◇

Because the logarithm arises frequently in our discussions in the next few sections, it is important to bear in mind some of the subtleties in its definition. Consider the function $\ln w$, with $w = Re^{j\theta}$. Here, $R > 0$ and θ is the phase of w. Clearly, replacing θ with $\theta + 2\pi k$ for integer k does not change the value of w. However, the value of $\ln w$ will depend on k. Thus,

$$\ln w = \text{Ln} R + j\theta + j2\pi k, \tag{6.4}$$

where Ln is the logarithm of a positive number as normally defined in calculus (Fig. 6.1). Thus, $\ln w$ has an infinite number of branches, each branch being defined by one value of the arbitrary integer k (Kreyszig, 1972; Churchill and Brown, 1984). If we restrict θ to be in the range $-\pi < \theta \leq \pi$ and set $k = 0$, then the logarithm $\ln w$ is said to be evaluated on the *principal branch* and denoted as $\text{Ln} w$ (called the *principal value*). For real and positive w, this agrees with Fig. 6.1. Note that $\text{Ln} 1 = 0$, but $\ln 1 = j2\pi k$, and its value depends on the branch number k.

Because the branch of the function $\ln w$ has not been specified in the statement of Lemma 6.1, the integer k is undetermined (and unimportant for our discussions). Similarly, the integer m is undetermined. However, in practically all our applications, we deal with quantities of the form Eq. (6.1), so that it is immaterial whether we use $\ln(.)$ or $\mathrm{Ln}(.)$ (because $\exp(j2\pi k) = 1$ for any integer k). The distinction should be kept in mind only when trying to follow detailed proofs.

Proof of Lemma 6.1. Note first that because $A(z)$ has no zeros on the unit circle, $\ln A(e^{j\omega})$ is well defined and finite there (except for branch ambiguity). We can express

$$A(e^{j\omega}) = \prod_{i=1}^{N}(1 - \alpha_i e^{-j\omega}), \tag{6.5}$$

so that

$$\ln A(e^{j\omega}) = \sum_{i=1}^{N} \ln(1 - \alpha_i e^{-j\omega}) + j2\pi k. \tag{6.6}$$

It is therefore sufficient to prove that

$$\frac{1}{2\pi} \int_0^{2\pi} \ln(1 - \alpha_i e^{-j\omega}) \, d\omega = j2\pi k_i, \tag{6.7}$$

where k_i is some integer. This proof depends crucially on the condition that the zeros of $A(z)$ are inside the unit circle, that is $|\alpha_i| < 1$. This condition enables us to expand the principal value $\mathrm{Ln}(1 - \alpha_i e^{-j\omega})$ using the power series[1]

$$\mathrm{Ln}(1 - v) = -\sum_{n=1}^{\infty} \frac{v^n}{n}, \quad |v| < 1. \tag{6.8}$$

Identifying v as $v = \alpha_i e^{-j\omega}$, we have

$$\ln(1 - \alpha_i e^{-j\omega}) = -\sum_{n=1}^{\infty} \frac{\alpha_i^n e^{-j\omega n}}{n} + j2\pi k_i. \tag{6.9}$$

The power series Eq. (6.8) has region of convergence $|v| < 1$. We know that any power series converges uniformly in its region of convergence and can therefore be integrated term by term (Kreyszig, 1972; Churchill and Brown, 1984). Because

$$\int_0^{2\pi} e^{-j\omega n} d\omega = 0, \quad n = 1, 2, 3\ldots, \tag{6.10}$$

---

[1]This expansion evaluates the logarithm on the principal branch because it reduces to zero when $v = 0$.

Eq. (6.7) readily follows, proving (6.2). Next, $A^*(e^{j\omega}) = \prod_{i=1}^{N}(1 - \alpha_i^* e^{j\omega})$. Because $|\alpha_i| < 1$, the above argument can be repeated to obtain a similar result. Thus,

$$\frac{1}{2\pi} \int_0^{2\pi} \ln|A(e^{j\omega})|^2 d\omega = \frac{1}{2\pi} \int_0^{2\pi} \ln A(e^{j\omega}) d\omega + \frac{1}{2\pi} \int_0^{2\pi} \ln A^*(e^{j\omega}) d\omega = j2\pi m$$

for some integer m. This completes the proof. $\qquad\square$

Derivation of Eq. (6.1). Recall that $e_N^f(n)$ is the output of the FIR filter $A_N(z)$, in response to the process $x(n)$. So, the power spectrum of $e_N^f(n)$ is given by $S_{xx}(e^{j\omega})|A_N(e^{j\omega})|^2$. Because $e_N^f(n)$ is white, its spectrum is constant, and its value at any frequency equals the mean square value $\mathcal{E}_N^f$ of $e_N^f(n)$. Thus,

$$\begin{aligned}
\mathcal{E}_N^f &= S_{xx}(e^{j\omega})|A_N(e^{j\omega})|^2 \\
&= \exp\left(\mathrm{Ln}\left[S_{xx}(e^{j\omega})|A_N(e^{j\omega})|^2 \right] \right) \\
&= \exp\left(\frac{1}{2\pi} \int_0^{2\pi} \mathrm{Ln}\left[S_{xx}(e^{j\omega})|A_N(e^{j\omega})|^2 \right] d\omega \right) \\
&= \exp\left(\frac{1}{2\pi} \int_0^{2\pi} \mathrm{Ln} S_{xx}(e^{j\omega}) \, d\omega + \frac{1}{2\pi} \int_0^{2\pi} \mathrm{Ln}|A_N(e^{j\omega})|^2 \, d\omega \right). \\
&= \exp\left(\frac{1}{2\pi} \int_0^{2\pi} \mathrm{Ln} S_{xx}(e^{j\omega}) \, d\omega \right),
\end{aligned}$$

because $\int_0^{2\pi} \mathrm{Ln}|A_N(e^{j\omega})|^2 \, d\omega = 0$ by Lemma 6.1. This establishes the desired result. The third equality above was possible because $S_{xx}(e^{j\omega})|A_N(e^{j\omega})|^2$ is constant for all ω.

6.3   A MEASURE OF SPECTRAL FLATNESS

Consider a zero-mean WSS process $x(n)$. We know that $x(n)$ is said to be white if its power spectrum $S_{xx}(e^{j\omega})$ is constant for all ω. When $x(n)$ is not white, it is of interest to introduce a "degree of whiteness" or "measure of flatness" for the spectrum. Although such a measure is by no means unique, the one introduced in this section proves to be useful in linear prediction theory, signal compression, and other signal processing applications.

Motivation from AM/GM ratio. The arithmetic-mean geometric mean inequality (abbreviated as the AM–GM inequality) says that any set of positive numbers $x_1, \ldots, x_N$ satisfies the inequality

$$\underbrace{\frac{1}{N}\sum_{i=1}^{N} x_i}_{\text{AM}} \geq \underbrace{\left(\prod_{i=1}^{N} x_i\right)^{1/N}}_{\text{GM}} \tag{6.11}$$

with equality if and only if all x_i are identical (Bellman, 1960). Thus,

$$0 \leq \frac{\text{GM}}{\text{AM}} \leq 1 \tag{6.12}$$

and the ratio GM/AM is a measure of how widely different the numbers $\{x_i\}$ are. As the numbers get closer to each other, this ratio grows and becomes unity when the numbers are identical. The ratio GM/AM can therefore be regarded as a measure of "flatness." The larger this ratio, the flatter the distribution is. □

By extending the concept of the GM/AM ratio to functions of a continuous variable (i.e., where the integer subscript i in x_i is replaced with a continuous variable, say frequency) we can obtain a useful measure of flatness for the power spectrum.

Definition 6.1. *Spectral flatness.* For a WSS process with power spectrum $S_{xx}(e^{j\omega})$, we define the *spectral flatness measure* as

$$\gamma_x^2 = \frac{\exp\left(\frac{1}{2\pi}\int_0^{2\pi} \mathrm{Ln} S_{xx}(e^{j\omega})\,d\omega\right)}{\frac{1}{2\pi}\int_0^{2\pi} S_{xx}(e^{j\omega})\,d\omega} \tag{6.13}$$

where it is assumed that $S_{xx}(e^{j\omega}) > 0$ for all ω. ◇

We will see that the ratio γ_x^2 can be regarded as the GM/AM ratio of the spectrum $S_{xx}(e^{j\omega})$. Clearly, $\gamma_x^2 > 0$. The denominator of γ_x^2 is

$$\frac{1}{2\pi}\int_0^{2\pi} S_{xx}(e^{j\omega})\,d\omega = R(0),$$

which is the mean square value of $x(n)$. We will first show that

$$\frac{1}{2\pi}\int_0^{2\pi} S_{xx}(e^{j\omega})\,d\omega \geq \exp\left(\frac{1}{2\pi}\int_0^{2\pi} \mathrm{Ln} S_{xx}(e^{j\omega})\,d\omega\right) \tag{6.14}$$

with equality if and only if $S_{xx}(e^{j\omega})$ is constant (i.e., $x(n)$ is white). Because of this result, we have

$$0 < \gamma_x^2 \leq 1,$$

with $\gamma_x^2 = 1$ if and only if $x(n)$ is white. A rigorous proof of a more general version of (6.14) can be found in Rudin (1974, pp. 63–64). The justification below is based on extending the AM–GM inequality for functions of continuous argument.

Justification of Eq. (6.14). Consider a function $f(y) > 0$ for $0 \leq y \leq a$. Suppose we divide the interval $0 \leq y \leq a$ into N uniform subintervals of length Δy as shown in Fig. 6.2. From the AM–GM inequality, we obtain

$$\frac{1}{N}\sum_{n=0}^{N-1} f(n\Delta y) \geq \prod_{n=0}^{N-1}\left[f(n\Delta y)\right]^{1/N}$$

$$= \exp\left(\text{Ln}\prod_{n=0}^{N-1}\left[f(n\Delta y)\right]^{1/N}\right)$$

$$= \exp\left(\frac{1}{N}\sum_{n=0}^{N-1}\text{Ln}\,f(n\Delta y)\right).$$

Equality holds if and only if $f(n\Delta y)$ is identical for all n. We can replace $1/N$ with $\Delta y/a$ (Fig. 6.2) so that the above becomes

$$\frac{1}{a}\sum_{n=0}^{N-1} f(n\Delta y)\Delta y \geq \exp\left(\frac{1}{a}\sum_{n=0}^{N-1}\text{Ln}\,f(n\Delta y)\Delta y\right).$$

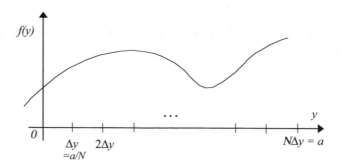

FIGURE 6.2: Pertaining to the proof of Eq. (6.14).

If we let $\Delta y \to 0$ (i.e., $N \to \infty$), the two summations in the above equation become integrals. We therefore have

$$\frac{1}{a} \int_0^a f(y) \, dy \geq \exp\left(\frac{1}{a} \int_0^a \text{Ln} f(y) \, dy\right). \qquad (6.15)$$

With $f(y)$ identified as $S_{xx}(e^{j\omega}) > 0$ and $a = 2\pi$, we immediately obtain Eq. (6.14). $\square$

Example 6.1. Spectral Flatness Measure. Consider a real, zero-mean WSS process $x(n)$ with the power spectrum $S_{xx}(e^{j\omega})$ shown in Fig. 6.3(a). Fig. 6.3(b) shows the spectrum for various values of α. It can be verified that

$$R(0) = \frac{1}{2\pi} \int_0^{2\pi} S_{xx}(e^{j\omega}) \, d\omega = 2.$$

The quantity in the numerator of Eq. (6.13) can be computed as

$$\exp\left(\frac{1}{2\pi} \int_0^{2\pi} \text{Ln} S_{xx}(e^{j\omega}) \, d\omega\right) = \exp\left(\alpha \text{Ln}\left(\frac{1}{\alpha} + 1\right)\right),$$

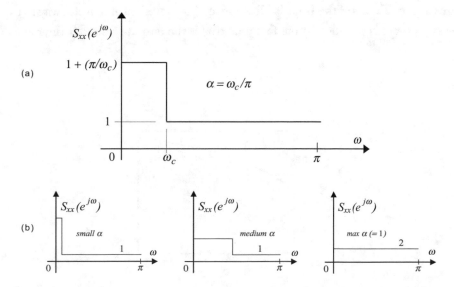

FIGURE 6.3: (a) Power spectrum for Ex. 6.1 and (b) three special cases of this power spectrum.

where $\alpha = \omega_c/\pi$. We can simplify this and substitute into the definition of γ_x^2 to obtain the flatness measure

$$\gamma_x^2 = 0.5 \left(1 + \frac{1}{\alpha} \right)^\alpha$$

This is plotted in Fig. 6.4 as a function of α. We see that as α increases from zero to unity, the spectrum gets flatter, as expected from an examination of Fig. 6.3(b).

6.4   SPECTRAL FLATNESS OF AN AR PROCESS

We know that if $x(n)$ is an AR(N) process, then the mean square value of the optimal prediction error is given by Eq. (6.1). The spectral flatness measure γ_x^2 defined in Eq. (6.13) therefore becomes

$$\gamma_x^2 = \frac{\mathcal{E}_N^f}{R(0)} \tag{6.16}$$

For fixed $R(0)$, the mean square prediction error $\mathcal{E}_N^f$ is smaller if the flatness measure γ_x^2 is smaller. Qualitatively speaking, this means that if $x(n)$ has a very nonflat power spectrum, then the prediction error $\mathcal{E}_N^f$ can be made very small. If on the other hand the spectrum $S_{xx}(e^{j\omega})$ of $x(n)$ is nearly flat, then the prediction error $\mathcal{E}_N^f$ is nearly as large as the mean square value of $x(n)$ itself.

Expression in Terms of the Impulse Response. An equivalent and useful way to express the above flatness measure is provided by the fact that $x(n)$ is the output of the IIR filter

$$G(z) = \frac{1}{A_N(z)},$$

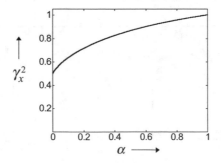

FIGURE 6.4: Flatness measure γ_x^2 in Ex. 6.1.

in response to the input $e_N^f(n)$ (Fig. 2.1(b)). Because $x(n)$ is $AR(N)$, the error $e_N^f(n)$ is white, and therefore, the power spectrum of $x(n)$ is

$$S_{xx}(e^{j\omega}) = \frac{\mathcal{E}_N^f}{|A_N(e^{j\omega})|^2}.$$

The mean square value of $x(n)$ is therefore given by

$$R(0) = \mathcal{E}_N^f \times \frac{1}{2\pi} \int_0^{2\pi} \frac{1}{|A_N(e^{j\omega})|^2} \, d\omega = \mathcal{E}_N^f \sum_{n=0}^{\infty} |g(n)|^2,$$

where $g(n)$ is the causal impulse response of $G(z)$. So, Eq. (6.16) becomes

$$\gamma_x^2 = \frac{1}{\sum_{n=0}^{\infty} |g(n)|^2} = \frac{1}{E_g},$$

where

$$E_g = \sum_{n=0}^{\infty} |g(n)|^2 = \text{ energy of the filter } G(z).$$

Summarizing, we have

Theorem 6.1. *Spectral flatness of AR process.* Let $x(n)$ be an $AR(N)$ process with mean square value $R(0) = E[|x(n)|^2]$. Let $A_N(z)$ be the optimal Nth-order prediction polynomial and $\mathcal{E}_N^f$ the corresponding mean square prediction error. Then, the spectral flatness γ_x^2 of the process $x(n)$ can be expressed in either of the following two forms:

1. $\gamma_x^2 = \mathcal{E}_N^f / R(0)$, or
2. $\gamma_x^2 = 1/\sum_{n=0}^{\infty} |g(n)|^2$,

where $g(n)$ is the causal impulse response of $G(z) = 1/A_N(z)$.   ◇

Example 6.2. Spectral Flatness of an AR(1) Process. In Ex. 3.3, we considered a process $x(n)$ with autocorrelation

$$R(k) = \rho^{|k|},$$

with $-1 < \rho < 1$. The optimal first-order prediction polynomial was found to be

$$A_1(z) = 1 - \rho z^{-1}.$$

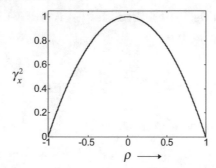

FIGURE 6.5: The flatness of an AR(1) process as a function of the correlation coefficient ρ.

We also saw that the first-order prediction error sequence $e_1^f(n)$ is white. This means that the process $x(n)$ is AR(1). The IIR all-pole filter $G(z)$ defined in Theorem 6.1 is

$$G(z) = \frac{1}{A_1(z)} = \frac{1}{1 - \rho z^{-1}}$$

and its impulse response is $g(n) = \rho^n \mathcal{U}(n)$. So $\sum_{n=0}^{\infty} g^2(n) = 1/(1 - \rho^2)$, and the flatness measure for the AR(1) process $x(n)$ is

$$\gamma_x^2 = 1 - \rho^2.$$

This is plotted in Fig. 6.5. So, the process gets flatter and flatter as ρ^2 gets smaller. The power spectrum of the process $x(n)$, which is the Fourier transform of $R(k)$, is given by

$$S_{xx}(e^{j\omega}) = \frac{1 - \rho^2}{|1 - \rho e^{-j\omega}|^2} = \frac{1 - \rho^2}{1 + \rho^2 - 2\rho \cos \omega}$$

This is depicted in Fig. 6.6 for two (extreme) values of ρ in the range $0 < \rho < 1$. (For $-1 < \rho < 0$, it is similar except that the peak occurs at π.) As ρ increases from zero to unity (i.e., the pole of

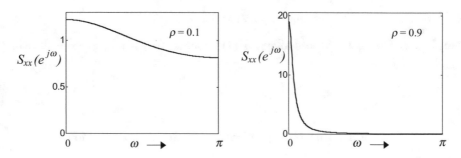

FIGURE 6.6: The power spectrum of an AR(1) process for two values of the correlation coefficient ρ.

$G(z)$ gets closer to the unit circle), the spectrum gets more and more peaky (less and less flat), consistent with the fact that γ_x^2 decreases as in Fig. 6.5.

6.5   CASE WHERE SIGNAL IS NOT AR

We showed that if $x(n)$ is an AR(N) process, then the Nth-order mean square prediction error is given by (6.1). When $x(n)$ is not AR, it turns out that this same quantity acts as a lower bound for the mean square prediction error as demonstrated in Fig. 6.7. We now prove this result.

Theorem 6.2. *Prediction error bound.* Let $x(n)$ be a WSS process with power spectrum $S_{xx}(e^{j\omega})$ finite and nonzero for all ω. Let $\mathcal{E}_m^f$ be the mean square value of the optimal mth-order prediction error. Then,

$$\mathcal{E}_m^f \geq \exp\left(\frac{1}{2\pi} \int_0^{2\pi} \mathrm{Ln} S_{xx}(e^{j\omega}) \, d\omega \right). \qquad (6.17)$$

Furthermore, equality holds if and only if $x(n)$ is AR(N) and $m \geq N$. $\diamond$

Proof. We know that the prediction error $e_m^f(n)$ is the output of the FIR filter $A_m(z)$ with input equal to $x(n)$ (Fig. 2.1(a)). So, its power spectrum is given by $S_{xx}(e^{j\omega})|A_m(e^{j\omega})|^2$, so that

$$\mathcal{E}_m^f = \frac{1}{2\pi} \int_0^{2\pi} S_{xx}(e^{j\omega})|A_m(e^{j\omega})|^2 \, d\omega$$

$$\geq \exp\left(\frac{1}{2\pi} \int_0^{2\pi} \mathrm{Ln}\left[S_{xx}(e^{j\omega})|A_m(e^{j\omega})|^2 \right] d\omega \right) \quad \text{(by Eq. (6.15))}$$

$$= \exp\left(\frac{1}{2\pi} \left[\int_0^{2\pi} \mathrm{Ln} S_{xx}(e^{j\omega}) \, d\omega + \int_0^{2\pi} \mathrm{Ln}|A_m(e^{j\omega})|^2 \, d\omega \right] \right)$$

$$= \exp\left(\frac{1}{2\pi} \int_0^{2\pi} \mathrm{Ln} S_{xx}(e^{j\omega}) \, d\omega \right) \quad \text{(by Lemma 6.1).}$$

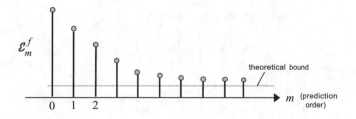

FIGURE 6.7: The optimal prediction error $\mathcal{E}_m^f$ decreases monotonically with increasing m but is lower bounded by Eq. (6.1) for any WSS process.

Equality holds if and only if $S_{xx}(e^{j\omega})|A_m(e^{j\omega})|^2$ is constant, or equivalently, $x(n)$ is AR with order $\leq m$. The proof is therefore complete. □

6.5.1 Error Spectrum Gets Flatter as Predictor Order Grows

We will now show that the power spectrum of the error $e_m^f(n)$ gets "flatter" as the prediction order m increases. With $S_m(e^{j\omega})$ denoting the power spectrum of $e_m^f(n)$, we have

$$S_m(e^{j\omega}) = S_{xx}(e^{j\omega})|A_m(e^{j\omega})|^2.$$

The flatness measure γ_m^2 for the process $e_m^f(n)$ is

$$\gamma_m^2 = \frac{\exp\left(\dfrac{1}{2\pi}\displaystyle\int_0^{2\pi} \text{Ln}S_m(e^{j\omega})\, d\omega\right)}{\dfrac{1}{2\pi}\displaystyle\int_0^{2\pi} S_m(e^{j\omega})\, d\omega}$$

$$= \frac{\exp\left(\dfrac{1}{2\pi}\displaystyle\int_0^{2\pi} \text{Ln}\left[S_{xx}(e^{j\omega})|A_m(e^{j\omega})|^2\right]\, d\omega\right)}{\mathcal{E}_m^f}$$

Using $\int_0^{2\pi} \text{Ln}|A_m(e^{j\omega})|^2\, d\omega = 0$ (Lemma 6.1), this simplifies to

$$\gamma_m^2 = \frac{\exp\left(\dfrac{1}{2\pi}\displaystyle\int_0^{2\pi} \text{Ln}S_{xx}(e^{j\omega})\, d\omega\right)}{\mathcal{E}_m^f}. \tag{6.18}$$

The numerator is fixed for a given process. As the prediction order increases, the error $\mathcal{E}_m$ decreases. This proves that the flatness is an increasing function of prediction order m (Fig. 6.8).

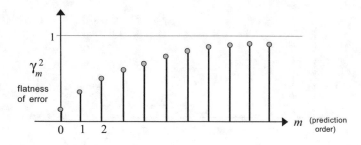

FIGURE 6.8: The flatness γ_m^2 of the optimal prediction error increases monotonically with increasing m. It is bounded above by unity.

We can rewrite γ_m^2 in terms of the flatness measure γ_x^2 of the original process $x(n)$ as follows:

$$\gamma_m^2 = \frac{\gamma_x^2 R(0)}{\mathcal{E}_m^f}.$$

Here, γ_x^2 and $R(0)$ are constants determined by the input process $x(n)$. The flatness γ_m^2 of the error process $e_m^f(n)$ is inversely proportional to the mean square value of the prediction error $\mathcal{E}_m^f$.

Example 6.3: Spectral Flattening With Linear Prediction. In this example, we consider an AR(4) process with power spectrum

$$S_{xx}(e^{j\omega}) = \frac{1}{|A(e^{j\omega})|^2},$$

where

$$A(z) = 1 - 1.8198z^{-1} + 2.0425z^{-2} - 1.2714z^{-3} + 0.4624z^{-4}$$

The first few elements of the autocorrelation are

$R(0)$	$R(1)$	$R(2)$	$R(3)$	$R(4)$	$R(5)$	$R(6)$	$R(7)$
7.674	4.967	−0.219	−3.083	−2.397	−0.639	−0.087	0.474

If we do linear prediction on this process, the prediction polynomials are

$A_0(z)$	1.0						
$A_1(z)$	1.0	−0.6473	0	0	0	0	0
$A_2(z)$	1.0	−1.1457	0.7701	0	0	0	0
$A_3(z)$	1.0	−1.5669	1.3967	−0.5469	0	0	0
$A_4(z)$	1.0	−1.8198	2.0425	−1.2714	0.4624	0	0
$A_5(z)$	1.0	−1.8198	2.0425	−1.2714	0.4624	0	0
$A_6(z)$	1.0	−1.8198	2.0425	−1.2714	0.4624	0	0

Thus, after $A_4(z)$, the polynomial does not change, because the original process is AR(4). The lattice coefficients are

k_1	k_2	k_3	k_4	k_5	k_6	k_7
−0.6473	0.7701	−0.5469	0.4624	0.0	0.0	0.0

Again, after k_4, all the coefficients are zero because the original process is AR(4).

Fig. 6.9 shows the power spectrum $S_m(e^{j\omega})$ of the prediction error $e_m^f(n)$ for $1 \leq m \leq 4$. The original power spectrum $S_{xx}(e^{j\omega}) = S_0(e^{j\omega})$ is shown in each plot for comparison. For convenience, the plots are shown normalized so that $S_{xx}(e^{j\omega})$ has peak value equal to unity. It is seen that the spectrum $S_m(e^{j\omega})$ gets flatter as m grows, and $S_4(e^{j\omega})$ is completely flat, again because the original process is AR(4). The flatness measure, calculated for each of these power spectra, is as follows:

$$\gamma_0^2 \qquad \gamma_1^2 \qquad \gamma_2^2 \qquad \gamma_3^2 \qquad \gamma_4^2 \quad \gamma_5^2 \quad \gamma_6^2$$

$$0.1309 \quad 0.2248 \quad 0.5517 \quad 0.7862 \quad 1.0 \quad 1.0 \quad 1.0$$

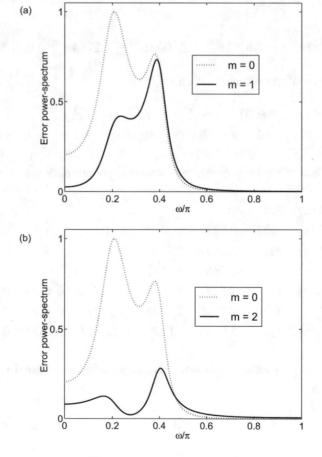

FIGURE 6.9: Original AR(4) power spectrum (dotted) and power spectrum of prediction error (solid). (a) Prediction order $= 1$, (b) prediction order $= 2$, (c) prediction order $= 3$, and (d) prediction order $= 4$.

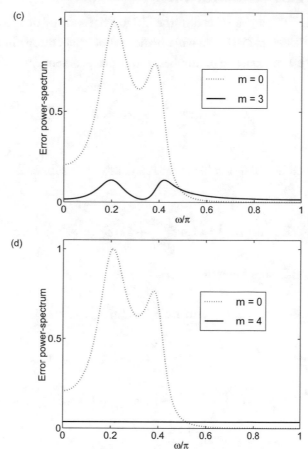

FIGURE 6.9: Continued.

6.5.2   Mean Square Error and Determinant

In Section 2.4.3, we found that the minimized mean square prediction error can be written in terms of the determinant of the autocorrelation matrix. Thus,

$$\det \mathbf{R}_m = \mathcal{E}_{m-1}^f \mathcal{E}_{m-2}^f \; \ldots \; \mathcal{E}_0^f. \qquad (6.19)$$

We know that the sequence $\mathcal{E}_m^f$ is monotone nonincreasing and lower bounded by zero. So it reaches a limit $\mathcal{E}_\infty^f$ as $m \to \infty$. We will now show that

$$\lim_{m \to \infty} \left(\det \mathbf{R}_m \right)^{1/m} = \mathcal{E}_\infty^f \qquad (6.20)$$

The quantity $(\det \mathbf{R}_m)^{1/m}$ can be related to the differential entropy of a Gaussian random process (Cover and Thomas, 1991, p. 273). We will say more about the entropy in Section 6.6.1.

Proof of (6.20). If a sequence of numbers $\alpha_0, \alpha_1, \alpha_2, \ldots$ tends to a limit α, then

$$\lim_{M \to \infty} \frac{1}{M} \sum_{i=0}^{M-1} \alpha_i = \alpha. \tag{6.21}$$

That is, the average value of the numbers tends to α as well (Problem 26). Defining $\alpha_i = \mathrm{Ln}\,\mathcal{E}_i^f$, we have from Eq. (6.19)

$$\mathrm{Ln}\left(\det \mathbf{R}_m\right) = \mathrm{Ln}\,\mathcal{E}_{m-1}^f + \mathrm{Ln}\,\mathcal{E}_{m-2}^f \ldots + \mathrm{Ln}\,\mathcal{E}_0^f.$$

Because $\mathrm{Ln}(x)$ is a continuous function when $x > 0$, we can say

$$\lim_{i \to \infty} \mathrm{Ln}\,\mathcal{E}_i^f = \mathrm{Ln}\,\mathcal{E}_\infty^f. \tag{6.22}$$

(Problem 26). We therefore have

$$\lim_{m \to \infty} \frac{1}{m} \mathrm{Ln}\left(\det \mathbf{R}_m\right) = \lim_{m \to \infty} \frac{1}{m}\left(\mathrm{Ln}\,\mathcal{E}_{m-1}^f + \mathrm{Ln}\,\mathcal{E}_{m-2}^f \ldots + \mathrm{Ln}\,\mathcal{E}_0^f\right)$$
$$= \mathrm{Ln}\,\mathcal{E}_\infty^f,$$

by invoking Eq. (6.21) and (6.22). This implies

$$\exp\left(\lim_{m \to \infty} \frac{1}{m} \mathrm{Ln}\left(\det \mathbf{R}_m\right)\right) = \exp\left(\mathrm{Ln}\,\mathcal{E}_\infty^f\right).$$

Because $\exp(x)$ is a continuous function of the argument x, we can interchange the limit with $\exp(.)$. This results in (6.20) indeed. □

A deeper result, called Szego's theorem (Haykin, 1986, pp. 96), can be used to prove the further fact that the limit in Eq. (6.20) is really equal to the bound on the right-hand side of Eq. (6.17). That is,

$$\lim_{m \to \infty}\left(\det \mathbf{R}_m\right)^{1/m} = \mathcal{E}_\infty^f = \exp\left(\frac{1}{2\pi} \int_0^{2\pi} \mathrm{Ln}\,S_{xx}(e^{j\omega})\,d\omega\right). \tag{6.23}$$

In other words, the lower bound on the mean square prediction error, which is achieved for the $AR(N)$ case for finite prediction order, is approached asymptotically for the non-AR case. For further details on the asymptotic behavior, see Grenander and Szego (1958) and Gray (1972).

Predictability measure. We now see that the flatness measure in Eq. (6.13) can be written in the form

$$\gamma_x^2 = \frac{\mathcal{E}_\infty^f}{R(0)} \qquad (6.24)$$

For very small γ_x^2, the power $\mathcal{E}_\infty^f$ in the prediction error is therefore much smaller than the power $R(0)$ in the original process, and we say that the process is very predictable. On the other hand, if γ_x^2 is close to unity then the ultimate prediction error is only slightly smaller that $R(0)$, and we say that the process is not very predictable. For white noise, $\gamma_x^2 = 1$, and the process is not predictable, as one would expect. The reciprocal $1/\gamma_x^2$ can be regarded as a convenient measure of predictability of a process. $\qquad \square$

6.6   MAXIMUM ENTROPY AND LINEAR PREDICTION

There is a close relation between linear prediction and the so-called maximum entropy methods (Burg, 1972). We now briefly outline this. In Section 6.5, we showed that the mth-order prediction error $\mathcal{E}_m^f$ for a WSS process $x(n)$ is bounded as

$$\mathcal{E}_m^f \geq \exp\left(\frac{1}{2\pi}\int_0^{2\pi} \text{Ln}S_{xx}(e^{j\omega})\, d\omega\right). \qquad (6.25)$$

We also know that if the process $x(n)$ is $AR(N)$, then equality is achieved in Eq. (6.25) whenever the prediction order $m \geq N$. Now consider two WSS processes $x(n)$ and $y(n)$ with autocorrelations $R(k)$ and $r(k)$, respectively, and power spectra $S_{xx}(e^{j\omega})$ and $S_{yy}(e^{j\omega})$, respectively. Assume that

$$R(k) = r(k), \quad |k| \leq N.$$

Then the predictor polynomials $A_m(z)$ and the mean square prediction errors $\mathcal{E}_m^f$ will be identical for the processes, for $1 \leq m \leq N$.

Suppose now that $y(n)$ is $AR(N)$, but $x(n)$ is not necessarily AR, then from Section 6.2, we know that

$$\exp\left(\frac{1}{2\pi}\int_0^{2\pi} \text{Ln}S_{yy}(e^{j\omega})\, d\omega\right) = \mathcal{E}_N^f$$

From Theorem 6.2, on the other hand, we know that

$$\mathcal{E}_N^f \geq \exp\left(\frac{1}{2\pi}\int_0^{2\pi} \mathrm{Ln}S_{xx}(e^{j\omega})\,d\omega\right)$$

Comparison of the two preceding equations shows that

$$\exp\left(\frac{1}{2\pi}\int_0^{2\pi} \mathrm{Ln}S_{yy}(e^{j\omega})\,d\omega\right) \geq \exp\left(\frac{1}{2\pi}\int_0^{2\pi} \mathrm{Ln}S_{xx}(e^{j\omega})\,d\omega\right).$$

Moreover, they are equal if and only if $x(n)$ is also AR(N) (by Theorem 6.2). Because $\exp(v)$ is a monotone-increasing function of real v, the above also means

$$\int_0^{2\pi} \mathrm{Ln}S_{yy}(e^{j\omega})d\omega \geq \int_0^{2\pi} \mathrm{Ln}S_{xx}(e^{j\omega})d\omega.$$

Summarizing, we have proved the following result.

Theorem 6.3. Let $x(n)$ and $y(n)$ be two WSS processes with autocorrelations $R(k)$ and $r(k)$, respectively, and power spectra $S_{xx}(e^{j\omega}) > 0$ and $S_{yy}(e^{j\omega}) > 0$, respectively. Assume that $r(k) = R(k)$ for $|k| \leq N$ and that $y(n)$ is AR(N), then

$$\underbrace{\int_0^{2\pi} \mathrm{Ln}S_{yy}(e^{j\omega})\,d\omega}_{AR(N)\ \mathrm{process}} \geq \int_0^{2\pi} \mathrm{Ln}S_{xx}(e^{j\omega})\,d\omega \tag{6.26}$$

with equality if and only if $x(n)$ is also AR. ◇

6.6.1 Connection to the Notion of Entropy

The importance of the above result lies in the fact that the integrals in Eq. (6.26) are related to the notion of entropy in information theory. If a WSS process $y(n)$ is Gaussian, then it can be shown (Cover and Thomas, 1991) that the quantity

$$\phi = a + b\int_0^{2\pi} \mathrm{Ln}S_{yy}(e^{j\omega})\,d\omega \tag{6.27}$$

is equal to the entropy per sample (entropy rate) of the process, where $a = \ln\sqrt{2\pi e}$ and $b = 1/4\pi$.

Details. For discrete-amplitude random variables, entropy is related to the extent of "uncertainty" or "randomness." For a continuous-amplitude random variable, the entropy is more appropriately called *differential* entropy and should be interpreted carefully. Thus, given a random variable x with differential entropy $\mathcal{H}_x$, consider the quantized version x_Δ with a fixed quantization step size Δ. Then the number of bits required to represent x_Δ is

$$b_\Delta = \mathcal{H}_x + c,$$

where c depends only on the step size Δ (Cover and Thomas, 1991, p. 229). Thus, for fixed step size, larger differential entropy implies more number of bits, uncertainty, or randomness.   □

Although the integral in the second term of (6.27) will be referred to as "entropy" and interpreted as "randomness" in the following discussions, the details given above should be kept in mind. The problem of maximizing ϕ under appropriate constraints is said to be an entropy maximization problem. Theorem 6.3 says that, among all Gaussian WSS processes with fixed autocorrelation for $|k| \le N$, the *entropy has the maximum value for AR(N) processes*. Furthermore, all AR(N) processes with autocorrelation fixed as above have the same entropy.

Now assume that $y(n)$ is the AR(N) model for $x(n)$ obtained by use of linear prediction as in Fig. 5.3. We know that in this model $e(n)$ is white noise with variance $\mathcal{E}_N^f$. If we also make $e(n)$ Gaussian, then the output $y(n)$ will be Gaussian as well (Papoulis, 1965). In other words, we have obtained an AR(N) model $y(n)$ for the process $x(n)$, with the following properties.

1. The AR(N) process $y(n)$ is Gaussian.
2. The autocorrelation $r(k)$ of the AR(N) process $y(n)$ matches the autocorrelation $R(k)$ of the given process $x(n)$ (i.e., $R(k) = r(k)$), for $|k| \le N$ (as shown in Section 5.4).
3. Under this matching constraint, the entropy $\int_0^{2\pi} \mathrm{Ln} S_{yy}(e^{j\omega})\, d\omega$ of the process $y(n)$ is the largest possible. In other words, $y(n)$ is "as random as it could be," under the constraint $r(k) = R(k)$ for $|k| \le N$.

The relation between linear prediction and maximum entropy methods is also discussed by Burg (1972), Robinson (1982), and Cover and Thomas (1991). Robinson (1982) also gives a good perspective on the history of spectrum estimation. Also see Papoulis (1981).

6.6.2   A Direct Maximization Problem

A second justification of Theorem 6.3 will now be mentioned primarily because it might be more insightful for some readers. Suppose we forget about the linear prediction problem and consider the following maximization problem directly: we are given a WSS process $x(n)$ with autocorrelation $R(k)$. We wish to find a WSS process $y(n)$ with the following properties.

1. Its autocorrelation $r(k)$ satisfies $r(k) = R(k)$ for $|k| \le N$.

2. With $S_{yy}(e^{j\omega})$ denoting the power spectrum of $y(n)$, the entropy $\int_0^{2\pi} \mathrm{Ln} S_{yy}(e^{j\omega})\, d\omega$ is as large as possible. This must be achieved by optimizing the coefficients $r(k)$ for $|k| > N$.

A necessary condition for the maximization[2] of the entropy ϕ defined in Eq. (6.27) is $\partial\phi/\partial r(k) = 0$ for $|k| > N$. Now,

$$\frac{\partial\phi}{\partial r(k)} = b \int_0^{2\pi} \frac{\partial[\mathrm{Ln}S_{yy}(e^{j\omega})]}{\partial r(k)}\, d\omega$$

$$= b \int_0^{2\pi} \frac{1}{S_{yy}(e^{j\omega})} \frac{\partial S_{yy}(e^{j\omega})}{\partial r(k)}\, d\omega$$

$$= b \int_0^{2\pi} \frac{1}{S_{yy}(e^{j\omega})} e^{-j\omega k}\, d\omega$$

The last equality follows by using $S_{yy}(e^{j\omega}) = \sum_{k=-\infty}^{\infty} r(k)e^{-j\omega k}$. Setting $\partial\phi/\partial r(k) = 0$, we therefore obtain

$$\int_0^{2\pi} \frac{1}{S_{yy}(e^{j\omega})} e^{-j\omega k}\, d\omega = 0, \qquad |k| > N.$$

This says that the inverse Fourier transform of $1/S_{yy}(e^{j\omega})$ should be of finite duration, restricted to the region $|k| \leq N$. In other words, we must be able to express $1/S_{yy}(e^{j\omega})$ as

$$\frac{1}{S_{yy}(e^{j\omega})} = c_1 |A_N(e^{j\omega})|^2,$$

or equivalently

$$S_{yy}(e^{j\omega}) = \frac{c}{|A_N(e^{j\omega})|^2},$$

where $A_N(z)$ is an FIR function of the form $A_N(z) = 1 + \sum_{i=1}^{N} a_{N,i}^* z^{-i}$. Thus, $S_{yy}(e^{j\omega})$ is an autoregressive spectrum.

Because the desired solution has been proved to be AR(N), we can find it simply by finding the AR(N) model of the process $x(n)$ using linear prediction techniques (Section 5.3). This is because we already know that the AR(N) model $y(n)$ arising out of linear prediction satisfies $r(k) = R(k)$ for $|k| \leq N$ (Section 5.4) and, furthermore, that an AR(N) spectrum satisfying the constraint $r(k) = R(k)$, $|k| \leq N$ is unique (because the solution to the normal equations is unique).

---

[2]The fact that this maximizes rather than minimizes ϕ is verified by examining the Hessian matrix (Chong and Žak, 2001; Antoniou and Lu, 2007).

6.6.3 Entropy of the Prediction Error

Let $e_m^f(n)$ be the mth-order optimal prediction error for a WSS process $x(n)$, and let $S_m(e^{j\omega})$ be the power spectrum for $e_m^f(n)$. With $S_{xx}(e^{j\omega})$ denoting the power spectrum of $x(n)$, we have

$$S_m(e^{j\omega}) = S_{xx}(e^{j\omega})|A_m(e^{j\omega})|^2$$

and $S_0(e^{j\omega}) = S_{xx}(e^{j\omega})$. Because $\int_0^{2\pi} \mathrm{Ln}|A_m(e^{j\omega})|^2 d\omega = 0$ (Lemma 6.1), we have

$$\int_0^{2\pi} \mathrm{Ln}S_m(e^{j\omega})\, d\omega = \int_0^{2\pi} \mathrm{Ln}S_{xx}(e^{j\omega})\, d\omega,$$

for all ω. If $x(n)$ is Gaussian, then so is $e_m^f(n)$, and the integral above represents the entropy of $e_m^f(n)$.

Thus, in the Gaussian case, the entropy of the prediction error is the same for all prediction orders m and equal to the entropy of the input $x(n)$. From Eq. (6.18), we see that the flatness γ_m^2 of the error $e_m^f(n)$ is given by

$$\gamma_m^2 = \frac{K}{\mathcal{E}_m^f}, \qquad (6.28)$$

where K is independent of m and depends only on the entropy of $x(n)$. So, the flatness of the prediction error is proportional to $1/\mathcal{E}_m^f$. Thus, as the prediction order m increases, the error spectrum becomes flatter, not because the entropy increases, but because the mean square error $\mathcal{E}_m^f$ decreases.

6.7 CONCLUDING REMARKS

The derivation of the optimal linear predictor was based only on the idea that the mean-squared prediction error should be minimized. We also pointed out earlier that the solution has the minimum-phase property and that it has structural interpretations in terms of lattices. We now see that the solution has interesting connections to flatness measures, entropy, and so forth. It is fascinating to see how all these connections arise starting from the "simple" objective of minimizing a mean square error.

Another interesting topic we have not discussed here is the prediction of a continuous-time *band-limited signal* from a finite number of past samples. With sufficiently large sampling rate, this can be done accurately, with predictor coefficients independent of the signal! This result dates back to Wainstein and Zubakov (1962). Further results can be found in Vaidyanathan (1987), along with a discussion of the history of this problem.

CHAPTER 7

Line Spectral Processes

7.1 INTRODUCTION

A WSS random process is said to be line spectral, if the power spectrum $S_{xx}(e^{j\omega})$ consists only of Dirac delta functions, that is,

$$S_{xx}(e^{j\omega}) = 2\pi \sum_{i=1}^{L} c_i \delta_a(\omega - \omega_i), \qquad 0 \le \omega < 2\pi, \tag{7.1}$$

where $\{\omega_i\}$ is a set of L distinct frequencies (called *line frequencies*) and $c_i \ge 0$. If $c_i > 0$ for all i, then L represents the number of lines in the spectrum as demonstrated in Fig. 7.1, and we say that L is the degree of the process. We also sometimes abbreviate this as a Linespec(L) process. Note that the power of the signal $x(n)$ around frequency ω_i is c_i, that is,

$$\frac{1}{2\pi} \int_{\omega_i - \epsilon}^{\omega_i + \epsilon} S_{xx}(e^{j\omega}) \, d\omega = c_i \tag{7.2}$$

for sufficiently small $\epsilon > 0$. So we say that c_i is the power at the line frequency ω_i.

In this chapter, we study line spectral processes in detail. We first express the Linespec(L) property concisely in terms of the autocorrelation matrix. We then study the time domain properties and descriptions of Linespec(L) processes. For example, we show that a Linespec(L) process satisfies a homogeneous difference equation. Using this, we can predict, with zero error, all of its samples, if we know the values of L successive samples. We then consider the problem of identifying a Linespec(L) process in noise. More specifically, suppose we have a signal $y(n)$, which is a sum of a Linespec(L) process $x(n)$ and uncorrelated white noise $e(n)$:

$$y(n) = x(n) + e(n).$$

We will show how the line frequencies ω_i and the line powers c_i in Eq. (7.1) can be extracted from the noisy signal.

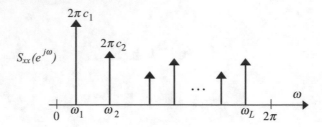

FIGURE 7.1: Power spectrum of a line spectral process with L lines. The kth line is a Dirac delta function with height $2\pi c_k$.

7.2   AUTOCORRELATION OF A LINE SPECTRAL PROCESS

The autocorrelation of the line spectral process is the inverse Fourier transform of Eq. (7.1):

$$R(k) = \sum_{i=1}^{L} c_i e^{j\omega_i k}. \tag{7.3}$$

This is a superposition of single-frequency sequences. Consider the $(L+1) \times (L+1)$ autocorrelation matrix $\mathbf{R}_{L+1}$ of a WSS process:

$$\mathbf{R}_{L+1} = \begin{bmatrix} R(0) & R(1) & \ldots & R(L) \\ R^*(1) & R(0) & \ldots & R(L-1) \\ \vdots & \vdots & \ddots & \vdots \\ R^*(L) & R^*(L-1) & \ldots & R(0) \end{bmatrix} \tag{7.4}$$

The Linespec(L) property can be stated entirely in terms of $\mathbf{R}_{L+1}$ and $\mathbf{R}_L$ as shown by the following result.

Theorem 7.1. *Line spectral processes.* A WSS process $x(n)$ is line spectral with degree L if and only if the matrix $\mathbf{R}_{L+1}$ is singular and $\mathbf{R}_L$ is nonsingular. $\diamond$

Proof. We already showed in Section 2.4.2 that if $\mathbf{R}_{L+1}$ is singular, then the power spectrum has the form Eq. (7.1) (line spectral, with degree $\leq L$). Conversely, assume $x(n)$ is line spectral with degree $\leq L$. If we pass $x(n)$ through an FIR filter $V(z)$ of order $\leq L$ with zeros at the line frequencies ω_i, the output is zero for all n. With the filter written as

$$V(z) = \sum_{i=0}^{L} v_i^* z^{-i}, \tag{7.5}$$

the output is

$$v_0^* x(n) + v_1^* x(n-1) + \ldots + v_L^* x(n-L) = 0. \tag{7.6}$$

Defining

$$\mathbf{x}(n) = \Big[x(n) \quad x(n-1) \ldots \quad x(n-L) \Big]^{\mathrm{T}}$$

and $\mathbf{v} = \Big[v_0 \ v_1 \ldots v_L \Big]^{\mathrm{T}}$, we can rewrite (7.6) as $\mathbf{v}^\dagger \mathbf{x}(n) = 0$. From this, we obtain $\mathbf{v}^\dagger E[\mathbf{x}(n)\mathbf{x}^\dagger(n)]\mathbf{v} = 0$, that is,

$$\mathbf{v}^\dagger \mathbf{R}_{L+1} \mathbf{v} = 0.$$

Because $\mathbf{R}_{L+1}$ is positive semidefinite, the above equation implies that $\mathbf{R}_{L+1}$ is singular. Summarizing, $\mathbf{R}_{L+1}$ is singular if and only if $x(n)$ is line spectral with degree $\leq L$. By replacing $L+1$ with L, it also follows that $\mathbf{R}_L$ is singular if and only if $x(n)$ is line spectral with degree $\leq L-1$. This means, in particular, that if $x(n)$ is line spectral with degree equal to L, then $\mathbf{R}_L$ has to be nonsingular and, of course, $\mathbf{R}_{L+1}$ singular.

Conversely, assume $\mathbf{R}_L$ is nonsingular and $\mathbf{R}_{L+1}$ singular. The latter implies that the process is line spectral with degree $\leq L$, as we already showed. The former implies that the degree is not $\leq L-1$. So the degree has to be exactly L. $\qquad\square$

Notice, by the way, that $\mathbf{R}_L$ is a principal submatrix of $\mathbf{R}_{L+1}$ in two ways:

$$\mathbf{R}_{L+1} = \begin{bmatrix} \mathbf{R}_L & \times \\ \times & R(0) \end{bmatrix} = \begin{bmatrix} R(0) & \times \\ \times & \mathbf{R}_L \end{bmatrix}. \tag{7.7}$$

We now show that if $x(n)$ is Linespec(L), there is a *unique* nonzero vector $\mathbf{v}$ (up to a scale factor) that annihilates $\mathbf{R}_{L+1}$:

$$\mathbf{R}_{L+1}\mathbf{v} = 0. \tag{7.8}$$

Proof. Because $\mathbf{R}_{L+1}$ is singular, it is obvious that there exists an annihilating vector $\mathbf{v}$. To prove uniqueness, we assume there are two linearly independent vectors $\mathbf{u} \neq 0$ and $\mathbf{w} \neq 0$, such that $\mathbf{R}_{L+1}\mathbf{u} = \mathbf{R}_{L+1}\mathbf{w} = 0$, and bring about a contradiction. The 0th elements of these vectors must be nonzero, that is, $u_0 \neq 0$ and $w_0 \neq 0$, for otherwise the remaining components annihilate $\mathbf{R}_L$ in view of (7.7), contradicting nonsingularity of $\mathbf{R}_L$. Now consider the vector $\mathbf{y} = \mathbf{u}/u_0 - \mathbf{w}/w_0$. This is nonzero (because $\mathbf{u}$ and $\mathbf{w}$ are linearly independent), but its zeroth component is zero:

$$\mathbf{y} = \begin{bmatrix} 0 \\ \mathbf{z} \end{bmatrix} \neq 0$$

Because $\mathbf{R}_{L+1}\mathbf{y} = 0$, it follows that $\mathbf{R}_L\mathbf{z} = 0, \quad \mathbf{z} \neq 0$, which violates nonsingularity of $\mathbf{R}_L$. $\qquad\square$

The result can also be proved by using the eigenvalue interlace property for principal submatrices of Hermitian matrices (Horn and Johnson, 1985). But the above proof is more direct.

7.2.1 The Characteristic Polynomial

The unique vector $\mathbf{v}$ that annihilates $\mathbf{R}_{L+1}$ has zeroth component $v_0 \neq 0$. For, if $v_0 = 0$, then the smaller vector

$$\mathbf{w} = \begin{bmatrix} v_1 \ v_2 \ \ldots \ v_L \end{bmatrix}^{\mathrm{T}}$$

would satisfy $\mathbf{R}_L \mathbf{w} = 0$ (in view of Eq. (7.7)) contradicting the fact that that $\mathbf{R}_L$ is nonsingular. So, without loss of generality, we can assume $v_0 = 1$. So the FIR filter $V(z)$ in Eq. (7.5), which annihilates the line spectral process $x(n)$, can be written as

$$V(z) = 1 + \sum_{i=1}^{L} v_i^* z^{-i}.$$

This filter, determined by the unique eigenvector $\mathbf{v}$, is itself unique and is called the *characteristic polynomial* or characteristic filter of the line spectral process. From Eq. (7.6), we see that a Linespec(L) process $x(n)$ also satisfies the *homogeneous difference equation*

$$x(n) = -\sum_{i=1}^{L} v_i^* x(n-i). \tag{7.9}$$

This means, in particular, that all future samples can be predicted with zero error if a block of L samples is known.

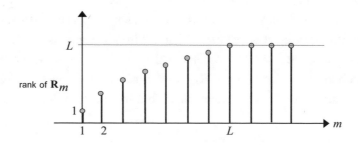

FIGURE 7.2: The rank of the autocorrelation matrix $\mathbf{R}_m$ as a function of size m, for a Linespec(L) process.

7.2.2 Rank Saturation for a Line Spectral Process

From the above discussions, we see that if $x(n)$ is Linespec(L), then $\mathbf{R}_L$ as well as $\mathbf{R}_{L+1}$ have rank L. It can, in fact, be proved (Problem 24) that

$$\text{rank } \mathbf{R}_m = \begin{cases} m & \text{for } 1 \le m \le L \\ L & \text{for } m \ge L. \end{cases} \tag{7.10}$$

This is indicated in Fig. 7.2. Thus, the rank *saturates* as soon as m exceeds the value L.

7.3 TIME DOMAIN DESCRIPTIONS

Line spectral processes exhibit a number of special properties, which are particularly useful when viewed in the time domain. In this section, we discuss some of these.

7.3.1 Extrapolation of Autocorrelation

We showed that a Linespec(L) process $x(n)$ sastisfies the homogeneous difference equation Eq. (7.9). If we multiply both sides of (7.9) by $x^*(n-k)$ and take expectations, we get

$$R(k) = -\sum_{i=1}^{L} v_i^* R(k-i), \tag{7.11}$$

where we have used $R(k) = E[x(n)x^*(n-k)]$. Conversely, we will see that whenever the autocorrelation satisfies the above recursion, the random process $x(n)$ itself satisfies the same recursion, that is, (7.9) holds.

Thus, if we know $R(k)$ for $0 \le k \le L-1$ for a Linespec(L) process $x(n)$, we can compute $R(k)$ for all k using (7.11). This is similar in spirit to the case where $x(n)$ is AR(N), in which case, the knowledge of $R(k)$ for $0 \le k \le N$ would reveal the polynomial $A_N(z)$ and therefore the value of $R(k)$ for all k (Section 5.3.2).

7.3.2 All Zeros on the Unit Circle

From the proof of Theorem 7.1, we know that $V(z)$ annihilates the process $x(n)$. This implies, in particular, that all the L zeros of

$$V(z) = 1 + \sum_{m=1}^{L} v_m^* z^{-m} \tag{7.12}$$

are on the unit circle at the L distinct points ω_i, where $S_{xx}(e^{j\omega})$ has the lines.[1] This means, in particular, that the numbers v_i satisfy the property

$$v_i = cv^*_{L-i}, \quad |c| = 1, \quad 0 \le i \le L, \tag{7.13}$$

where $v_0 = 1$. That is, $V(z)$ is a linear phase filter. Because $V(z)$ has the L distinct zeros $e^{j\omega_i}$, the solutions to the homogeneous equation Eq. (7.9) necessarily have the form

$$x(n) = \sum_{i=1}^{L} d_i e^{j\omega_i n}. \tag{7.14}$$

Note that there are only L random variables d_i in Eq. (7.14), although the "random" process $x(n)$ is an infinite sequence. The WSS property of $x(n)$ imposes some strong conditions on the joint behavior of the random variables d_i, which we shall derive soon. The deterministic quantity c_i in Eq. (7.3) is the *power* of the process $x(n)$ at the line frequency ω_i, whereas the random variable d_i is said to be the random *amplitude* (possibly complex) at the frequency ω_i.

7.3.3   Periodicity of a Line Spectral Process

The process (7.14) is periodic if and only if all the frequencies ω_i are rational multiples of 2π (Oppenheim and Schafer, 1999). That is,

$$\omega_i = \left(\frac{K_i}{M_i}\right) 2\pi$$

for integers K_i, M_i. In this case, we can write $\omega_i = 2\pi L_i/M$ where M is the least common multiple (LCM) of the set $\{M_i\}$, and L_i are integers. Thus, all the frequencies are harmonics of the fundamental frequency $2\pi/M$. Under this condition, we say that the process is *periodic* with period M or just *harmonic*. Because $R(k)$ has the same form as $x(n)$ (compare Eqs. (7.3) and (7.14)), we see that $R(k)$ is periodic if and only if $x(n)$ is periodic. In this connection, the following result is particularly interesting.

Theorem 7.2. Let $x(n)$ be a WSS random process with autocorrelation $R(k)$. Suppose $R(0) = R(M)$ for some $M > 0$. Then, $x(n)$ is periodic with period M, and so is $R(k)$.   ◇

   Proof. Because $R(0) = R(M)$, we have

$$E[x(n)x^*(n)] = E[x(n)x^*(n-M)]. \tag{7.15}$$

---

[1]Another way to see this would be as follows: $R(k)$ satisfies the homogeneous difference equation (7.11). At the same time, it also has the form (7.3). From the theory of homogeneous equations, it then follows that the L zeros of $V(z)$ have the form $e^{j\omega_i}$.

The mean square value of $x(n) - x(n - M)$, namely,

$$E\left[\left(x(n) - x(n - M)\right)\left(x^*(n) - x^*(n - M)\right)\right],$$

can be written as

$$\mu = E[|x(n)|^2] - E[x(n)x^*(n - M)]$$
$$-E[x(n - M)x^*(n)] + E[|x(n - M)|^2].$$

The first and the last terms are equal because of WSS property. Substituting further from Eq. (7.15), we get $\mu = 0$. So the random variable $x(n) - x(n - M)$ has mean square value equal to zero. This implies $x(n) = x(n - M)$. So $x(n)$ is periodic with period M. Using the definition $R(k) = E[x(n)x^*(n - k)]$, it is readily verified that $R(k)$ is also periodic, with period M.   □

7.3.4   Determining the Parameters of a Line Spectral Process

Given the autocorrelation $R(k)$, $0 \le k \le L$ of the Linespec(L) process $x(n)$, we can identify the unique eigenvector $\mathbf{v}$ satisfying $\mathbf{R}_{L+1}\mathbf{v} = 0$. Then, the unique characteristic polynomial $V(z)$ is determined (Eq. (7.12)). All its L zeros are guaranteed to be on the unit circle (i.e., they have the form $e^{j\omega_i}$). The line frequencies ω_i can therefore be identified. From Eq. (7.3), we can solve for the constants c_i (line powers) by writing a set of linear equations. For example, if $L = 3$, we have

$$\begin{bmatrix} 1 & 1 & 1 \\ e^{j\omega_1} & e^{j\omega_2} & e^{j\omega_3} \\ e^{2j\omega_1} & e^{2j\omega_2} & e^{2j\omega_3} \end{bmatrix} \begin{bmatrix} c_1 \\ c_2 \\ c_3 \end{bmatrix} = \begin{bmatrix} R(0) \\ R(1) \\ R(2) \end{bmatrix}. \tag{7.16}$$

The 3×3 matrix is a Vandermonde matrix and is nonsingular because ω_i are distinct (Horn and Johnson, 1985). So, we can uniquely identify the line powers c_i. Thus, the line frequencies and powers in Fig. 7.1 have been completely determined from the set of coefficients $R(k)$, $0 \le k \le L$.

Identifying amplitudes of sinusoids in $x(n)$. Similarly, consider the samples $x(n)$ of the Linespec(L) random process $x(n)$. If we are given the values of L successive samples, say $x(0), x(1), \ldots x(L - 1)$, we can uniquely identify the coefficients d_i appearing in Eq. (7.14) by solving a set of equations similar to (7.16). Thus, once we have measured L samples of the random process, we can find all the future and past samples exactly, with zero error! The randomness is only in the L coefficients $\{d_i\}$, and once these have been identified, $x(n)$ is known for all n.   □

7.4   FURTHER PROPERTIES OF TIME DOMAIN DESCRIPTIONS

Starting from the assumption that $x(n)$ is a line spectral process, we derived the time domain expression Eq. (7.14). Conversely, consider a sequence of the form Eq. (7.14), where d_i are random variables. Is it necessarily a line spectral process? Before addressing this question, we first have to address WSS property. The WSS property of Eq. (7.14) itself imposes some severe restrictions on the random variables d_i, as shown next.

Lemma 7.1. Consider a random process of the form

$$x(n) = \sum_{i=1}^{L} d_i e^{j\omega_i n},$$

(7.17)

where d_i are random variables and ω_i are distinct constants. Then, $x(n)$ is a WSS process if and only if

1. $E[d_i] = 0$ whenever $\omega_i \neq 0$ (zero-mean condition).
2. $E[d_i d_m^*] = 0$ for $i \neq m$ (orthogonality condition).   ◇

Proof. Recall that $x(n)$ is WSS if $E[x(n)]$ and $E[x(n)x^*(n-k)]$ are independent of n. First, assume that the two conditions in the lemma are satisfied. We have $E[x(n)] = \sum_{i=1}^{L} E[d_i] e^{j\omega_i n}$. By the first condition of the lemma, this is constant for all n. Next,

$$E[x(n)x^*(n-k)] = \sum_{i=1}^{L} \sum_{m=1}^{L} E[d_i d_m^*] e^{j\omega_i n} e^{-j\omega_m (n-k)}$$

(7.18)

With d_i satisfying the second condition of the lemma, this reduces to

$$E[x(n)x^*(n-k)] = \sum_{i=1}^{L} E[|d_i|^2] e^{j\omega_i k},$$

(7.19)

which is independent of n. Thus, the two conditions of the lemma imply that $x(n)$ is WSS. Now consider the converse. Suppose $x(n)$ is WSS. Because $E[x(n)]$ is independent of n, we see from Eq. (7.17) that $E[d_i] = 0$ whenever $\omega_i \neq 0$. So the first condition of the lemma is necessary. Next, Eq. (7.18) can be rewritten as

$$E[x(n)x^*(n-k)] = \sum_{i=1}^{L} \sum_{m=1}^{L} e^{j\omega_i n} E[d_i d_m^*] e^{-j\omega_m n} e^{j\omega_m k}.$$

Express the right-hand side as

$$
\underbrace{\big[e^{j\omega_1 n} \,\ldots\, e^{j\omega_L n}\big]\mathbf{D}
\begin{bmatrix}
e^{-j\omega_1 n} & 0 & \cdots & 0 \\
0 & e^{-j\omega_2 n} & \cdots & 0 \\
\vdots & \vdots & \ddots & \vdots \\
0 & 0 & \cdots & e^{-j\omega_L n}
\end{bmatrix}}_{\text{call this } \mathbf{A}(n)}
\underbrace{\begin{bmatrix}
e^{j\omega_1 k} \\
e^{j\omega_2 k} \\
\vdots \\
e^{j\omega_L k}
\end{bmatrix}}_{\mathbf{t}(k)},
$$

where $\mathbf{D}$ is an $L \times L$ matrix with $[\mathbf{D}]_{im} = E[d_i d_m^*]$. WSS property requires that $\mathbf{A}(n)\mathbf{t}(k)$ be independent of n, for *any* fixed choice of k. Thus, defining the matrix

$$
\mathbf{T} = \big[\,\mathbf{t}(0)\ \mathbf{t}(1) \,\ldots\, \mathbf{t}(L-1)\,\big],
$$

we see that $\mathbf{A}(n)\mathbf{T}$ must be independent of n. But $\mathbf{T}$ is a Vandermonde matrix and is nonsingular because ω_i are distinct. So $\mathbf{A}(n)$ must itself be independent of n. Consider now its mth column:

$$
[\mathbf{A}(n)]_m = \sum_{i=1}^{L} E[d_i d_m^*]e^{j(\omega_i - \omega_m)n}.
$$

Because ω_i are distinct, the differences $\omega_i - \omega_m$ are distinct for fixed m. So the above sum is independent of n if and only if $E[d_i d_m^*] = 0$ for $i \neq m$. The second condition of the lemma is, therefore, necessary as well. □

From Eq. (7.19), it follows that $R(k)$ has the form

$$
R(k) = \sum_{i=1}^{L} E[|d_i|^2]e^{j\omega_i k} \tag{7.20}
$$

so that the power spectrum is a line spectrum. Thus, if d_i satisfies the conditions of the lemma, then $x(n)$ is not only WSS, but is also a line spectral process of degree L. Summarizing, we have proved this:

Theorem 7.3. Consider a random process of the form

$$
x(n) = \sum_{i=1}^{L} d_i e^{j\omega_i n}, \tag{7.21}
$$

where d_i are random variables and ω_i are distinct constants. Then, $x(n)$ is a WSS and, hence, a line spectral process if and only if

1. $E[d_i] = 0$ whenever $\omega_i \neq 0$ (zero-mean condition).
2. $E[d_i d_m^*] = 0$ for $i \neq m$ (orthogonality condition).

Under this condition, $R(k)$ is as in Eq. (7.20), so that $E[|d_i|^2]$ represents the power at the line frequency ω_i. ◇

Example 7.1: Stationarity and Ergodicity of Line spectral Processes. Consider the random process $x(n) = Ae^{j\omega_0 n}$, where $\omega_0 \neq 0$ is a constant and A is a zero-mean random variable. We see that $E[x(n)] = E[A]e^{j\omega_0 n} = 0$ and $E[x(n)x^*(n-k)] = E[|A|^2]e^{j\omega_0 k}$. Because these are independent of n, the process $x(n)$ is WSS.

If a process $x(n)$ is ergodic (Papoulis, 1965), then, in particular, $E[|x(n)|^2]$ must be equal to the time average. In our case, because $|x(n)| = |A|$, we see that

$$\frac{1}{2N+1} \sum_{n=-N}^{N} |x(n)|^2 = |A|^2$$

and is, in general, a random variable! So the time average will in general not be equal to the ensemble average $E[|A|^2]$, unless $|A|^2$ is nonrandom. (For example, if $A = ce^{j\theta}$, where c is a constant and θ is a random variable, then $|A|^2$ would be a constant.) So we see that, a WSS line spectral process is not ergodic unless we impose some strong restriction on the random amplitudes.

Example 7.2: Real Random Sinusoid. Next consider a real random process of the form

$$x(n) = A\cos(\omega_0 n + \theta),$$

where A and θ are real random variables and $\omega_0 > 0$ is a constant. We have

$$x(n) = 0.5Ae^{j\theta}e^{j\omega_0 n} + 0.5Ae^{-j\theta}e^{-j\omega_0 n}.$$

From Lemma 7.1, we know that this is WSS if and only if

$$E[Ae^{j\theta}] = 0, \quad E[Ae^{-j\theta}] = 0, \quad \text{and} \quad E[A^2 e^{j2\theta}] = 0. \qquad (7.22)$$

This is satisfied, for example, if A and θ are statistically independent random variables and θ is uniformly distributed in $[0, 2\pi)$. Thus, statistical independence implies

$$E[Ae^{j\theta}] = E[A]E[e^{j\theta}] \quad \text{and} \quad E[A^2 e^{j2\theta}] = E[A^2]E[e^{j2\theta}],$$

and the uniform distribution of θ implies $E[e^{j\theta}] = E[e^{j2\theta}] = 0$, so that Eq. (7.22) is satisfied. Notice that Lemma 7.1 requires that $Ae^{j\theta}$ and $Ae^{-j\theta}$ have zero mean for WSS property, but it is not necessary for A itself to have zero mean! Summarizing, $x(n) = A\cos(\omega_0 n + \theta)$ is WSS whenever the real random variables A and θ are statistically independent and θ is uniformly distributed in $[0, 2\pi]$. It is easily verified that $E[x(n)] = 0$ and

$$R(k) = E[x(n)x(n-k)] = 0.5E[A^2]\cos(\omega_0 k) = P\cos(\omega_0 k),$$

where $P = 0.5E[A^2]$. So the power spectrum is

$$S_{xx}(e^{j\omega}) = \frac{P}{2} \times 2\pi \left(\delta_a(\omega - \omega_0) + \delta_a(\omega + \omega_0) \right), \ 0 \leq \omega < 2\pi.$$

The quantity P is the total power in the two line frequencies $\pm\omega_0$.

Example 7.3: Sum of Sinusoids. Consider a real random process of the form

$$x(n) = \sum_{i=1}^{N} A_i \cos(\omega_i n + \theta_i). \tag{7.23}$$

Here A_i and θ_i are real random variables and ω_i are distinct positive constants. Suppose we make the following statistical assumptions:

1. A_i and θ_m are statistically independent for any i, m.
2. θ_i and θ_m are statistically independent for $i \neq m$.
3. θ_m is uniformly distributed in $0 \leq \theta_m < 2\pi$ for any m.

We can then show that $x(n)$ is WSS. For this, we only have to rewrite $x(n)$ in the complex form Eq. (7.21) and verify that the conditions of Theorem 7.3 are satisfied (Problem 25). Notice that no assumption has been made about the joint statistics of A_i and A_m, for $i \neq m$. Under the above conditions, it can be verified that $E[x(n)] = 0$. How about the autocorrelation? We have

$$\begin{aligned} R(k) &= E[x(n)x(n-k)] \\ &= E[x(0)x(-k)] \quad \text{(using WSS)} \\ &= E \sum_{i=1}^{N} A_i \cos(\theta_i) \sum_{m=1}^{N} A_m \cos(-k\omega_m + \theta_m). \end{aligned}$$

Using the assumptions listed above, the cross terms for $i \neq m$ reduce to

$$E[A_i A_m]E[\cos(\theta_i)]E[\cos(-k\omega_m + \theta_m)] = 0.$$

When $i = m$, the terms reduce to $P_i \cos(\omega_i k)$ as in Ex. 7.2. Thus,

$$R(k) = E[x(n)x(n-k)] = \sum_{i=1}^{N} P_i \cos(\omega_i k),$$

where $P_i = 0.5E[A_i^2] > 0$. The power spectrum is therefore

$$S_{xx}(e^{j\omega}) = \pi \sum_{i=1}^{N} P_i \left(\delta_a(\omega - \omega_i) + \delta_a(\omega + \omega_i) \right), \quad 0 \leq \omega < 2\pi.$$

The total power at the frequencies ω_i and $-\omega_i$ is given by $0.5P_i + 0.5P_i = P_i$.

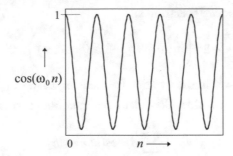

FIGURE 7.3: Plot of the cosine.

Philosophical Discussions. A random process of the form Eq. (7.21), where there are only L random variables d_i, always appears to hold some conceptual mystery. If we measure the L random variables d_i by some means, then we know the entire waveform $x(n)$. So the samples are fully predictable, as also seen from the difference equation Eq. (7.9). Furthermore, a plot of $x(n)$ does not look "random." For example, consider $A\cos(\omega_0 n + \theta)$ discussed in Ex. 7.2. Here, the random variables A and θ do not depend on n, and the plot of $x(n)$ as a function of n (Fig. 7.3) is just a nice and smooth cosine!

So, where does the randomness manifest? Recall that a random process, by definition, is a collection (or ensemble) of waveforms (Papoulis, 1965; Peebles, 1987). Each waveform is an outcome of an experiment. For example, the result of an experiment determines the values of the random variables A and θ to be A_0 and θ_0, and the outcome $A_0\cos(\omega_0 n + \theta_0)$ is fully determined for all n and is said to be a realization of the random process. It is for this reason that it is somewhat tricky to force ergodicity (Ex. 7.1); ergodicity says that time averages are equal to ensemble averages, but if a particular outcome (time function) of the random process does not exhibit "randomness" along time, then the time average cannot tell us much about the ensemble averages indeed.

7.5   PREDICTION POLYNOMIAL OF LINE SPECTRAL PROCESSES

Let $x(n)$ be Linespec(L) and let $\mathbf{R}_{L+1}$ be its autocorrelation matrix of size $(L+1) \times (L+1)$. In Section 7.2, we argued that there exists a unique vector $\mathbf{v}$ of the form

$$\mathbf{v} = \begin{bmatrix} 1 & v_1 & \dots & v_L \end{bmatrix}^{\mathrm{T}}$$

such that $\mathbf{R}_{L+1}\mathbf{v} = 0$. By comparison with the augmented normal equation Eq. (2.20), we see that if we take the coefficients of the Lth-order predictor to be

$$a_{L,i} = v_i,$$

then the optimal Lth-order prediction error $\mathcal{E}_L^f$ becomes zero. The optimal predictor polynomial for the process $x(n)$ is therefore

$$A_L(z) = 1 + v_1^* z^{-1} + \ldots + v_L^* z^{-L} = V(z).$$

In Section 7.3.2, we saw that $V(z)$ has all the L zeros on the unit circle, at the points $e^{j\omega_i}$, where ω_i are the line frequencies. We therefore conclude that the optimal Lth-order predictor polynomial $A_L(z)$ of a Linespec(L) process has all L zeros on the unit circle. This also implies

$$a_n = c a_{L-n}^*$$

for some c with $|c| = 1$. For example, in the real case, a_n is symmetrical or antisymmetrical. In what follows, we present two related results.

Lemma 7.2. Let $R(k)$ be the autocorrelation of a WSS process $x(n)$, and let $R(k)$ satisfy the difference equation

$$R(k) = -\sum_{i=1}^{L} a_i^* R(k-i). \qquad (7.24)$$

Furthermore, let there be no such equation of smaller degree satisfied by $R(k)$. Then the process $x(n)$ is line spectral with degree L and the polynomial $A_L(z) = 1 + \sum_{i=1}^{L} a_i^* z^{-i}$ necessarily has all the L zeros on the unit circle. $\diamond$

Proof. By using the property $R(k) = R^*(-k)$, we can verify that Eq. (7.24) implies

$$\underbrace{\begin{bmatrix} R(0) & R(1) & \ldots & R(L) \\ R^*(1) & R(0) & \ldots R(L-1) \\ \vdots & \vdots & \ddots & \vdots \\ R^*(L) & R^*(L-1) \ldots & R(0) \end{bmatrix}}_{\mathbf{R}_{L+1}} \begin{bmatrix} 1 \\ a_L \\ \vdots \\ a_L \end{bmatrix} = 0.$$

So $\mathbf{R}_{L+1}$ is singular. If $\mathbf{R}_L$ were singular, we could proceed as in Section 7.2.1 to show that $x(n)$ satisfies a difference equation of the form Eq. (7.9) (but with lower-degree $L-1$); from this, we could derive an equation of the form Eq. (7.24) but with lower degree $L-1$. Because this violates the conditions of the lemma, we conclude that $\mathbf{R}_L$ is nonsingular. Using Theorem 7.1, we therefore conclude that the process $x(n)$ is Linespec(L). It is clear from Section 7.3 that the polynomial $A_L(z)$ is the characteristic polynomial $V(z)$ of the process $x(n)$ and therefore has all zeros on the unit circle. $\square$

We now present a related result directly in terms of the samples $x(n)$ of random process. This is a somewhat surprising conclusion, although its proof is very simple in light of the above discussions. (For a more direct proof, see Problem 23.)

Theorem 7.4. Let $x(n)$ be a random process satisfying

$$x(n) = -\sum_{i=1}^{L} a_i^* x(n-i). \qquad (7.25)$$

Furthermore, let there be *no such equation of smaller degree* satisfied by $x(n)$. If the process $x(n)$ is WSS, then

1. The polynomial $A_L(z) = 1 + \sum_{i=1}^{L} a_i^* z^{-i}$ has all the L zeros on the unit circle.
2. $x(n)$ is line spectral with degree L.   ◇

Thus, if a process satisfies the difference equation Eq. (7.25) (and if L is the smallest such number), then imposing the WSS property on $x(n)$ immediately forces the zeros of $A_L(z)$ to be all on the unit circle!

Proof. Suppose the process is WSS. If we multiply both sides of Eq. (7.25) by $x^*(n-k)$ and take expectations, this results in Eq. (7.24). If there existed an equation of the form Eq. (7.24) with L replaced by $L-1$, we could prove then that $\mathbf{R}_L$ is singular and proceed as in Section 7.2.1 to derive an equation of the form Eq. (7.25) with L replaced by $L-1$. Because this violates the conditions of the theorem, there is no smaller equation of the form Eq. (7.24). Applying Lemma 7.2, the claim of the theorem immediately follows.   □

7.6   SUMMARY OF PROPERTIES

Before we proceed to line spectral processes buried in noise, we summarize, at the expense of being somewhat repetitive, the main properties of line spectral processes.

1. *Spectrum has only impulses.* We say that a WSS random process is Linespec(L) if the power spectrum is made of L impulses, that is,

$$S_{xx}(e^{j\omega}) = 2\pi \sum_{i=1}^{L} c_i \delta_a(\omega - \omega_i), \qquad 0 \le \omega < 2\pi, \qquad (7.26)$$

with $c_i > 0$ for any i. Here, ω_i are distinct and called the line frequencies and c_i are the powers at these line frequencies. In this case, the process $x(n)$ as well as its autocorrelation

$R(k)$ satisfy an Lth-order recursive, homogeneous, difference equation:

$$x(n) = -\sum_{i=1}^{L} v_i^* x(n-i), \quad R(k) = -\sum_{i=1}^{L} v_i^* R(k-i). \tag{7.27}$$

So all the samples of the sequence $x(n)$ can be predicted with zero error, if L successive samples are known. Similarly, all the autocorrelation coefficients $R(k)$ of the process $x(n)$ can be computed if $R(k)$ is known for $0 \leq k \leq L-1$.

2. *Characteristic polynomial.* In the above, v_i^* are the coefficients of the polynomial

$$V(z) = 1 + \sum_{i=1}^{L} v_i^* z^{-i}. \tag{7.28}$$

This is called the characteristic polynomial of the Linespec(L) process $x(n)$. This polynomial has all its L zeros on the unit circle at $z = e^{j\omega_i}$.

3. *Sum of sinusoids.* The random process $x(n)$ as well as the autocorrelation $R(k)$ can be expressed as a sum of L single-frequency signals

$$x(n) = \sum_{i=1}^{L} d_i e^{j\omega_i n}, \quad R(k) = \sum_{i=1}^{L} c_i e^{j\omega_i k}, \tag{7.29}$$

where d_i are random variables (amplitudes at the line frequencies ω_i) and $c_i = E[|d_i|^2]$ (*powers* at the line frequencies ω_i).

4. *Autocorrelation matrix.* A WSS process $x(n)$ is Linespec(L) if and only if the autocorrelation matrix $\mathbf{R}_{L+1}$ is singular and $\mathbf{R}_L$ nonsingular. So $\mathbf{R}_{L+1}$ has rank L. There is a unique vector $\mathbf{v}$ of the form

$$\mathbf{v} = \begin{bmatrix} 1 & v_1 & \dots & v_L \end{bmatrix}^{\mathrm{T}} \tag{7.30}$$

satisfying $\mathbf{R}_{L+1}\mathbf{v} = 0$, and this vector uniquely determines the characteristic polynomial $V(z)$. Furthermore, the rank has the saturation behavior sketched in Fig. 7.2.

5. *Sum of sinusoids, complex.* A random process of the form $x(n) = \sum_{i=1}^{L} d_i e^{j\omega_i n}$ is WSS and, hence, line spectral, if and only if (a) $E[d_i] = 0$ whenever $\omega_i \neq 0$ and (b) $E[d_i d_m^*] = 0$ for $i \neq m$. Under this condition, $R(k) = \sum_{i=1}^{L} E[|d_i|^2] e^{j\omega_i k}$ (Theorem 7.3).

6. *Sum of sinusoids, real.* Consider the process $x(n) = \sum_{i=1}^{N} A_i \cos(\omega_i n + \theta_i)$, where A_i and θ_i are real random variables and ω_i are distinct positive constants. This is WSS, hence, line spectral, under the conditions stated in Ex. 7.3. The autocorrelation is then $R(k) = \sum_{i=1}^{N} P_i \cos(\omega_i k)$, where $P_i = 0.5 E[A_i^2]$ is the total power in the line frequencies $\pm\omega_i$.

7. *Identifying the parameters.* Given the set of autocorrelation coefficients $R(k)$, $0 \leq k \leq L$ of the Linespec(L) process $x(n)$, we can determine v_i in Eq. (7.28) from the eigenvector $\mathbf{v}$ of $\mathbf{R}_{L+1}$ corresponding to zero eigenvalue. The line frequencies ω_i can then be determined because $e^{j\omega_i}$ are the zeros of the characteristic polynomial $V(z)$. Once the line frequencies are known, the constants c_i are determined as described in Section 7.3.4.

8. *Periodicity.* If a WSS process $x(n)$ is such that $R(0) = R(M)$ for some $M > 0$, then $x(n)$ is periodic with period M, and so is $R(k)$ (Theorem 7.2). So the process is line spectral.

9. *Prediction polynomial.* If we perform optimal linear prediction on a Linespec(L) process, we will find that the Lth-order prediction polynomial $A_L(z)$ is equal to the characteristic polynomial $V(z)$ and therefore has all its zeros on the unit circle. The prediction error $\mathcal{E}_L^f = 0$.

7.7   IDENTIFYING A LINE SPECTRAL PROCESS IN NOISE

Identification of sinusoids in noise is an important problem in signal processing. It is also related to the problem of finding the direction of arrival of a propagating wave using an array of sensors. An excellent treatment can be found in Therrien (1992). We will be content with giving a brief introduction to this problem here. Consider a random process

$$y(n) = x(n) + e(n), \qquad (7.31)$$

where $x(n)$ is Linespec(L) (i.e., a line spectral process with degree L). Let $e(n)$ be zero-mean white noise with variance σ_e^2, that is,

$$E[e(n)e^*(n-k)] = \sigma_e^2 \delta(k).$$

We thus have a line spectral process buried in noise (popularly referred to as *sinusoids buried in noise*). The power spectrum of $x(n)$ has line frequencies (impulses), whereas that of $e(n)$ is flat, with value σ_e^2 for all frequencies. The total power spectrum is demonstrated in Fig. 7.4. Notice that some of the line frequencies could be very closely spaced, and some could be buried beneath the noise level.

Assume that $x(n)$ and $e(m)$ are uncorrelated, that is,

$$E[x(n)e^*(m)] = 0, \qquad \text{for any } n, m.$$

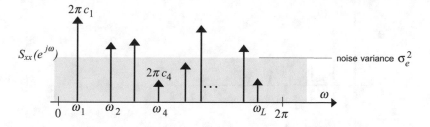

FIGURE 7.4: A Linespec(L) process with additive white noise of variance σ_e^2.

Let $r(k)$ and $R(k)$ denote the autocorrelation sequences of $y(n)$ and $x(n)$, respectively. Using this and the whiteness of $e(n)$, we get

$$
\begin{aligned}
r(k) &= E[y(n)y^*(n-k)] \\
&= E[x(n)x^*(n-k)] + E[e(n)e^*(n-k)] \\
&= R(k) + \sigma_e^2\delta(k),
\end{aligned}
$$

which yields

$$r(k) = R(k) + \sigma_e^2\delta(k) \tag{7.32}$$

Note, in particular, that $r(k) = R(k), k \neq 0$. Thus, the $m \times m$ autocorrelation matrix $\mathbf{r}_m$ of the process $y(n)$ is given by

$$\mathbf{r}_m = \mathbf{R}_m + \sigma_e^2\mathbf{I}_m, \tag{7.33}$$

where $\mathbf{R}_m$ is the $m \times m$ autocorrelation matrix of $x(n)$. We say that $\mathbf{R}_m$ is the signal autocorrelation matrix and $\mathbf{r}_m$ is the signal-plus-noise autocorrelation matrix.

7.7.1 Eigenstructure of the Autocorrelation Matrix

Let η_i be an eigenvalue of $\mathbf{R}_m$ with eigenvector $\mathbf{u}_i$, so that

$$\mathbf{R}_m\mathbf{u}_i = \eta_i\mathbf{u}_i.$$

Then

$$\underbrace{\left[\mathbf{R}_m + \sigma_e^2\mathbf{I}_m\right]}_{\mathbf{r}_m}\mathbf{u}_i = [\eta_i + \sigma_e^2]\mathbf{u}_i.$$

Thus, the eigenvalues of $\mathbf{r}_m$ are

$$\lambda_i = \eta_i + \sigma_e^2.$$

The corresponding eigenvectors of $\mathbf{r}_m$ are the same as those of $\mathbf{R}_m$.

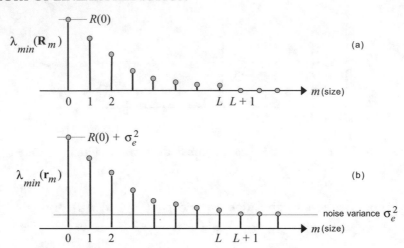

FIGURE 7.5: (a) The decreasing minimum-eigenvalue of the autocorrelation matrix $\mathbf{R}_m$ as the size m increases and (b) corresponding behavior of the minimum eigenvalue of $\mathbf{r}_m$.

The Minimum Eigenvalue. Because $x(n)$ is Linespec(L), its autocorrelation matrix $\mathbf{R}_m$ is nonsingular for $1 \le m \le L$ and singular for $m = L+1$ (Theorem 7.1). Thus, the smallest eigenvalue of $\mathbf{R}_m$ is positive for $m \le L$ and equals zero for $m > L$. It is easily shown (Problem 12) that the smallest eigenvalue $\lambda_{\min}(\mathbf{R}_m)$ of the matrix $\mathbf{R}_m$ cannot increase with m. So it behaves in a monotone manner as depicted in Fig. 7.5(a).

Accordingly, the smallest eigenvalue of $\mathbf{r}_m$ behaves as in Fig. 7.5(b). In particular, for $m > L$ it attains a constant value equal to σ_e^2. We can use this as a means of identifying the number of line frequencies L in the process $x(n)$, as well as the variance σ_e^2 of the noise $e(n)$. (This could be misleading in some cases, as we cannot keep measuring the eigenvalue for an unlimited number of values of m; see Problem 29.)

Eigenvector corresponding to the smallest eigenvalue. Let $\mathbf{v}$ be the eigenvector of $\mathbf{r}_{L+1}$ corresponding to the smallest eigenvalue σ_e^2. So

$$\mathbf{r}_{L+1}\mathbf{v} = \sigma_e^2 \mathbf{v}. \tag{7.34}$$

In view of Eq. (7.33) this implies

$$\mathbf{R}_{L+1}\mathbf{v} + \sigma_e^2 \mathbf{v} = \sigma_e^2 \mathbf{v},$$

that is,

$$\mathbf{R}_{L+1}\mathbf{v} = 0.$$

Thus, $\mathbf{v}$ is also the eigenvector that annihilates $\mathbf{R}_{L+1}$. The eigenvector corresponding to the minimum eigenvalue of $\mathbf{r}_{L+1}$ can be computed using any standard method, such as, for example, the power method (Golub and Van Loan, 1989). Once we identify the vector $\mathbf{v}$, then we can find the characteristic polynomial $V(z)$ (Eq. (7.28)). Its roots are guaranteed to be on the unit circle (i.e., have the form $e^{j\omega_i}$), revealing the line frequencies. The smallest eigenvalue of $\mathbf{r}_{L+1}$ gives the noise variance σ_e^2. □

7.7.2   Computing the Powers at the Line Frequencies

The autocorrelation $R(k)$ of the process $x(n)$ has the form (7.3), where c_i are the powers at the line frequencies ω_i. Because we already know ω_i, we can compute c_i by solving linear equations (as in Eq. (7.16)), if we know enough samples of $R(k)$. For this note that $R(k)$ is related to the given autocorrelation $r(k)$ as in Eq. (7.32). So we know all the samples of $R(k)$ (because σ_e^2 has been identified as described above).

What if we do not know the noise variance? Even if the noise variance σ_e^2 and, hence, $R(0)$ are not known, we can compute the line powers as follows. From Eq. (7.32) we have $R(k) = r(k)$, $k \neq 0$. So we know the values of $R(1), \ldots, R(L)$. We can write a set of L linear equations

$$
\begin{bmatrix}
e^{j\omega_1} & e^{j\omega_2} & \ldots & e^{j\omega_L} \\
e^{j2\omega_1} & e^{j2\omega_2} & \ldots & e^{j2\omega_L} \\
\vdots & \vdots & \ddots & \vdots \\
e^{jL\omega_1} & e^{jL\omega_2} & \ldots & e^{jL\omega_L}
\end{bmatrix}
\begin{bmatrix}
c_1 \\ c_2 \\ \vdots \\ c_L
\end{bmatrix}
=
\begin{bmatrix}
R(1) \\ R(2) \\ \vdots \\ R(L)
\end{bmatrix}
\tag{7.35}
$$

The $L \times L$ matrix above is Vandermonde and nonsingular because the line frequencies ω_i are distinct (Horn and Johnson, 1985). So we can solve for the line powers c_i uniquely. □

The technique described in this section to estimate the parameters of a line spectral process is commonly referred to as *Pisarenko's harmonic decomposition*. Notice that the line frequencies need not be harmonically related for the method to work (i.e., ω_k need not be integer multiples of a fundamental ω_0).

Case of Real Processes. The case where $x(n)$ is a real line spectral process is of considerable importance. From Ex. 7.3 recall that a process of the form

$$
x(n) = \sum_{i=1}^{N} A_i \cos(\omega_i n + \theta_i)
$$

is WSS and hence, harmonic, under some conditions. If $x(n)$ in Eq. (7.31) has this form, then the above discussions continue to hold. We can identify the eigenvector $\mathbf{v}$ of $\mathbf{r}_{L+1}$ corresponding to its minimum eigenvalue (Eq. (7.34)) and, hence, the corresponding line frequencies as described previously. In this case, it is more convenient to write $R(k)$ in the form

$$R(k) = \sum_{i=1}^{N} P_i \cos(\omega_i k),$$

where $P_i > 0$ is the total power at the frequencies $\pm \omega_i$. Because $R(k)$ is known for $1 \leq k \leq N$, we can identify P_i by writing linear equations, as we did for the complex case.

Example 7.4: Estimation of sinusoids in noise. Consider the sum of sinusoids

$$x(n) = 0.3 \sin(0.1\pi n) + 1.6 \sin(0.5\pi n) + 2.1 \sin(0.9\pi n), \tag{7.36}$$

and assume that we have noisy samples

$$y(n) = x(n) + e(n),$$

where $e(n)$ is zero-mean white noise with variance $\sigma_e^2 = 0.1$. We assume that 100 samples of $y(n)$ are available. The signal component $x(n)$ and the noise component $e(n)$ generated using the Matlab command *randn* are shown in Fig. 7.6. The noisy data $y(n)$ can be regarded as a process with a line spectrum (six lines because each sinusoid has two lines), buried in a background noise spectrum (as in Fig. 7.4). We need the 7×7 autocorrelation matrix $\mathbf{R}_7$ of the process $y(n)$, to estimate the lines. So, from the noisy data $y(n)$, we estimate the autocorrelation $R(k)$ for $0 \leq k \leq 6$ using the *autocorrelation* method described in Section 2.5. This estimates the top row of $\mathbf{R}_7$:

$$\begin{bmatrix} 3.6917 & -2.0986 & 0.5995 & -1.4002 & 2.1572 & -0.1149 & -1.8904 \end{bmatrix}$$

From these coefficients, we form the 7×7 Toeplitz matrix $\mathbf{R}_7$ and compute its smallest eigenvalue and the corresponding eigenvector

$$\mathbf{v} = \begin{bmatrix} 0.6065 & 0.0083 & -0.3619 & 0.0461 & -0.3619 & 0.0083 & 0.6065 \end{bmatrix}^{\mathrm{T}}.$$

With the elements of $\mathbf{v}$ regarded as the coefficients of the polynomial $V(z)$, we now compute the zeros of $V(z)$ and obtain

$$0.0068 \pm 0.9999j, \quad 0.9414 \pm 0.3373j, \quad -0.9550 \pm 0.2965j.$$

From the angles of these zeros, we identify the six line frequencies $\pm\omega_1$, $\pm\omega_2$, and $\pm\omega_3$. The three sinusoidal frequencies ω_1, ω_2, and ω_3 are estimated to be

$$0.1029\pi, \quad 0.4985\pi, \quad \text{and} \quad 0.9034\pi,$$

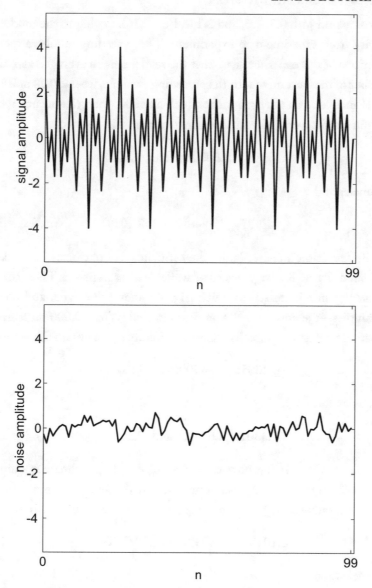

FIGURE 7.6: Example 7.4. Estimation of sinusoids in noise. The plots show the signal and the noise components used in the example.

which are quite close to the correct values given in Eq. (7.36), namely, 0.1π, 0.5π, and 0.9π. With ω_k thus estimated and $R(k)$ already estimated, we can use (7.35) to estimate c_k and finally obtain the amplitudes of the sinusoids (positive square roots of c_k in this example). The result is

$$0.3058, \quad 1.6258, \quad \text{and} \quad 2.1120,$$

which should be compared with 0.3, 1.6, and 2.1 in Eq. (7.36). Owing to the randomness of noise, these estimates vary from experiment to experiment. The preceding numbers represent averages over several repetitions of this experiment, so that the reader sees an averaged estimate.

Large number of measurements. In the preceding example, we had 100 noisy measurements available. If this is increased to a large number such as 5000, then the accuracy of the estimate improves. The result (again averaged over several experiments) is

$$0.1001\pi, \quad 0.5000\pi, \quad \text{and} \quad 0.9001\pi$$

for the line frequencies and

$$0.2977, \quad 1.5948, \quad \text{and} \quad 2.1019$$

for the line amplitudes. If the noise is large, the estimates will obviously be inaccurate, but this effect can be combated if there is a large record of available measurements. Thus, consider the above example with noise variance increased to unity. Fig. 7.7 shows the signal and noise components for the first few samples—the noise is very significant indeed. With 10,000 measured samples $y(n)$ used in the estimation process (an unrealistically large number), we obtain the estimate

$$0.1005\pi, \quad 0.4999\pi, \quad 0.9000\pi,$$

for the line frequencies and

$$0.3301, \quad 1.6140, \quad \text{and} \quad 2.1037$$

for the line amplitudes. Considering how large the noise is, this estimate is quite good indeed. Figure 7.8 shows even larger noise, almost comparable with the signal (noise variance $\sigma_e^2 = 2$). With 10,000 measured samples $y(n)$ used in the estimation process , we obtain the estimate

$$0.1030\pi, \quad 0.5039\pi \quad \text{and} \quad 0.9010\pi$$

for the line frequencies and

$$0.2808, \quad 1.6162, \quad \text{and} \quad 2.0725$$

for the line amplitudes.   □

In this section, we have demonstrated that sinusoids buried in additive noise can be identified effectively from the zeros of an eigenvector of the estimated autocorrelation. The method outlined above reveals the fundamental principles as advanced originally by Pisarenko (1973). Since then, many improved methods have been proposed, which can obtain more accurate estimates from a small

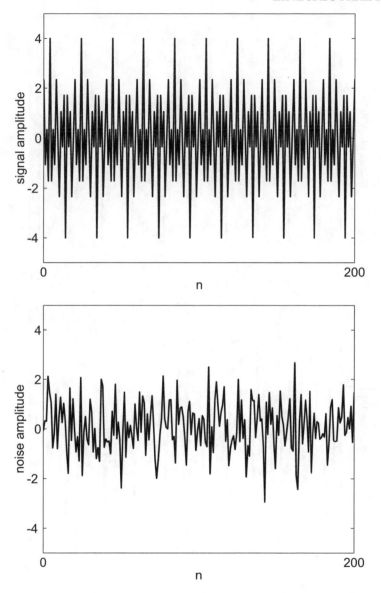

FIGURE 7.7: Example 7.4. Estimation of sinusoids in large noise. The plots show the signal and the noise components used in the example.

number of noisy measurements. Notable among these are *minimum-norm* methods (Kumaresan and Tufts, 1983), *MUSIC* (Schmidt, 1979, 1986), and *ESPRIT* (Paulraj et al., 1986; Roy and Kailath, 1989). These methods can also be applied to the estimation of direction of arrival information in

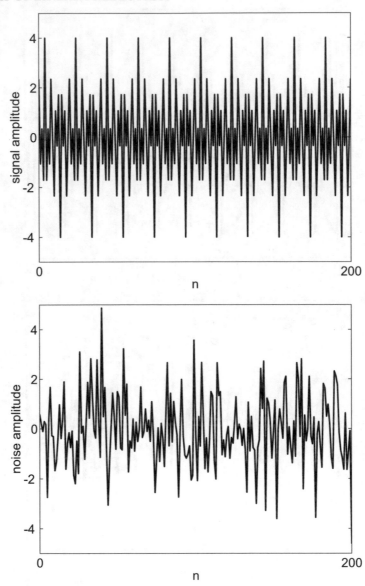

FIGURE 7.8: Example 7.4. Estimation of sinusoids in very large noise. The plots show the signal and the noise components used in the example.

array processing applications (Van Trees, 2002). In fact, many of these ideas originated in the array processing literature (Schmidt, 1979). A review of these methods can be found in the very readable account given by Therrien (1992).

7.8   LINE SPECTRUM PAIRS

In Section 5.6, we discussed the application of linear prediction theory in signal compression. We explained the usefulness of the lattice coefficients $\{k_m\}$ in the process. Although the quantization and transmission of lattice coefficients $\{k_m\}$ guarantees that the reconstruction filter $1/A_N(z)$ remains stable, the coefficients k_m are difficult to interpret from a perceptual viewpoint. If we knew the relative importance of various coefficients k_m in preserving perceptual quality, we could assign them bits according to that. But this has not been found to be possible.

In speech coding practice, one uses an important variation of the prediction coefficients called *line-spectrum pairs* (LSP). The motivation comes from the fact that the connection between LSP and perceptual properties is better understood (Itakura, 1975; Soong and Juang, 1984, 1993; Kang and Fransen, 1995). The set of LSP coefficients not only guarantees stability of the reconstruction filter under quantization, but, in addition, a better perceptual interpretation in the frequency domain is obtained. To define the LSP coefficients, we construct two new polynomials

$$P(z) = A_N(z) + z^{-(N+1)}\widetilde{A}_N(z),$$

$$Q(z) = A_N(z) - z^{-(N+1)}\widetilde{A}_N(z). \qquad (7.37)$$

Notice that these can be regarded as the polynomial $A_{N+1}(z)$ in Levinson's recursion for the cases $k_{N+1} = 1$ and $k_{N+1} = -1$, respectively. Equivalently, they are the transfer functions of the FIR lattice shown in Fig. 7.9.

Let us examine the zeros of these polynomials. The zeros of $P(z)$ and $Q(z)$ are solutions of the equations

$$z^{-(N+1)}\left(\frac{\widetilde{A}_N(z)}{A_N(z)}\right) = -1 = e^{j(2m+1)\pi} \quad [\text{for } P(z)]$$

$$z^{-(N+1)}\left(\frac{\widetilde{A}_N(z)}{A_N(z)}\right) = 1 = e^{j2m\pi} \quad [\text{for } Q(z)].$$

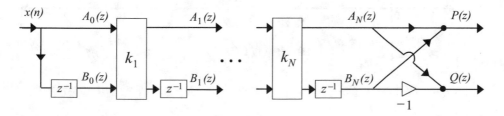

FIGURE 7.9: Interpretation of the LSP polynomials $P(z)$ and $Q(z)$ in terms of the FIR LPC lattice.

Because the left-hand side is all-pass of order $N+1$ with all zeros in $|z| < 1$, it satisfies the *modulus property*

$$\left| \frac{z^{-(N+1)}\widetilde{A}_N(z)}{A_N(z)} \right| \begin{cases} < 1 \ \text{for} |z| > 1 \\ > 1 \ \text{for} |z| < 1 \\ = 1 \ \text{for} |z| = 1 \end{cases}$$

See Vaidyanathan (1993, p. 75) for a proof. Thus, the solutions of the preceding equations are necessarily on the unit circle. That is, all the $N+1$ zeros of $P(z)$ and similarly those of $Q(z)$ are on the unit circle.

Alternation Property. In fact, we can say something deeper. It is well known (Vaidyanathan, 1993, p. 76) that the phase response $\phi(\omega)$ of the stable all-pass filter

$$\frac{z^{-(N+1)}\widetilde{A}_N(z)}{A_N(z)}$$

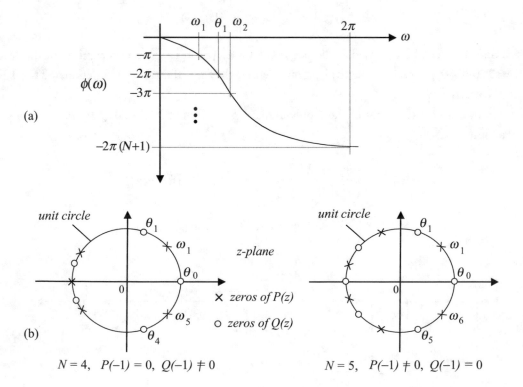

FIGURE 7.10: Development of LSPs. (a) Monotone phase response of $z^{-(N+1)}\widetilde{A}_N(z)/A_N(z)$ and (b) zeros of $P(z)$ and $Q(z)$ for the cases $N = 4$ and $N = 5$.

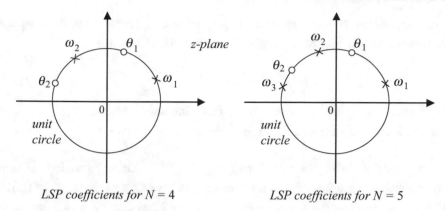

LSP coefficients for N = 4   LSP coefficients for N = 5

FIGURE 7.11: Examples of LSP coefficients for $N = 4$ and $N = 5$. The unit-circle coefficients in the lower half are redundant and therefore dropped.

is a monotone decreasing function, spanning a total range of $2(N + 1)\pi$ (Fig. 7.10(a)). Thus, the $N + 1$ zeros of $P(z)$ and $Q(z)$ *alternate* with each other as demonstrated in Fig. 7.10(b) for $N = 4$ and $N = 5$. The angles of these zeros are denoted as ω_i and θ_i.

Because $A_N(z)$ has real coefficients in speech applications, the zeros come in complex conjugate pairs. Note that $P(-1) = 0$ for even N and $Q(-1) = 0$ for odd N. Moreover, $Q(1) = 0$ and $P(1) \neq 0$ regardless of N. These follow from the fact that $A_N(\pm 1) = \tilde{A}_N(\pm 1)$. Thus, if we know the zeros of $P(z)$ and $Q(z)$ in the upper half of the unit circle (and excluding $z = \pm 1$), we can fully determine these polynomials (their constant coefficients being unity by the definition (7.37)). There are N such zeros (ω_k and θ_k), as demonstrated in Fig. 7.11 for $N = 4$ and $N = 5$. These are called *line spectrum pairs* (LSP) associated with the predictor polynomial $A_N(z)$. For example, the LSP parameters for $N = 4$ are the four ordered frequencies

$$\omega_1 < \theta_1 < \omega_2 < \theta_2 \qquad (7.38)$$

and the LSP parameters for $N = 5$ are the five ordered frequencies

$$\omega_1 < \theta_1 < \omega_2 < \theta_2 < \omega_3. \qquad (7.39)$$

Properties and Advantages of LSPs. Given the N coefficients of the polynomial $A_N(z)$, the N LSP parameters can be uniquely identified by finding the zeros of $P(z)$ and $Q(z)$. These parameters are quantized, making sure that the ordering (e.g., Eq. (7.39)) is preserved in the process. The quantized coefficients are then transmitted. Because $P(z)$ and $Q(z)$ can be computed uniquely from the LSP parameters, approximations of these polynomial can be computed at the receiver. From these, we can identify an approximation of $A_N(z)$ because

$$A_N(z) = (P(z) + Q(z))/2.$$

The speech segment can then be reconstructed from the stable filter $1/A_N(z)$ as described in Section 5.6. Several features of the LSP coding scheme are worth noting.

1. *Stability preserved*. As long as the ordering (e.g., Eq. (7.39)) is preserved in the quantization, the reconstructed version of $A_N(z)$ is guaranteed to have all zeros in $|z| < 1$. Thus, stability of $1/A_N(z)$ is preserved despite quantization as in the case of lattice coefficient quantization.

2. *Perceptual spectral interpretation*. Unlike the lattice coefficients, the LSP coefficients are perceptually better understood. To explain this, recall first that for sufficiently large N, the $AR(N)$ model gives a good approximation of the speech power spectrum $S_{xx}(e^{j\omega})$. This approximation is especially good near the peak frequencies, called the *formants* of speech. Now, the peak locations correspond approximately to the pole angles of the filter $1/A_N(z)$. Near these locations, the phase response tends to change rapidly. The same is true of the phase response $\phi(\omega)$ of the all-pass filter, as shown in Fig. 7.12(b). The LSP coefficients, which are intersections of the horizontal lines (multiples of π) with the plot of $\phi(\omega)$, therefore tend to get *crowded* near the formant frequencies. Thus, the crucial features of the power spectrum tend to get coded into the *density-information* of the LSP coefficients at various points on the unit circle. This information allows us to perform bit allocation among the LSP coefficients (quantize the crowded LSP coefficients with greater accuracy). For a given bit rate, this results in perceptually better speech quality, as compared with the quantization of lattice coefficients. Equivalently, for a given perceptual quality, we can reduce the bit rate; in early work, an approximately 25% saving has been reported using this idea, and more savings have been obtained in a series of other papers. Furthermore, as the bit rate is reduced, the speech degradation is found to be more gradual compared with lattice coefficient quantization.

3. *Acoustical tube models*. It has been shown in speech literature that the lattice structure is related to an acoustical tube model of the vocal tract (Markel and Gray, 1976). This is the origin of the term *reflection coefficients*. The values $k_{N+1} = \pm 1$ in Fig. 7.9 indicate, for example, situations where one end of the tube is open or closed. Thus, the LSP frequencies are related to the open- and close-ended acoustic tube models.

4. *Connection to circuit theory*. In the theory of passive electrical circuits, the concept of *reactances* is well-known (Balabanian and Bickart, 1969). Reactances are input impedances of electrical LC networks. It turns out that reactances have a pole-zero alternation property similar to the

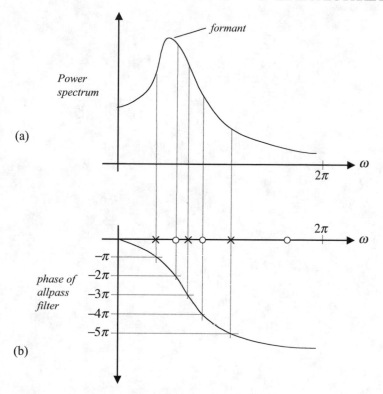

FIGURE 7.12: Explanation of how the LSP coefficients tend to get crowded near the formant regions of the speech spectrum. (a) A toy power spectrum with one formant and (b) the phase response of the all-pass filter $z^{-(N+1)}\widetilde{A}_N(z)/A_N(z)$, where $A_N(z)$ is the prediction polynomial (see text).

alternation of the zeros of the LSP polynomials $P(z)$ and $Q(z)$. In fact, the ratio $P(z)/Q(z)$ is nothing but a discrete time version of the reactance function.

7.9   CONCLUDING REMARKS

In this chapter, we presented a rather detailed study of line spectral processes. The theory finds applications in the identification of sinusoids buried in noise. A variation of this problem is immediately applicable in array signal processing where one seeks to identify the direction of arrival of a signal using an antenna array (Van Trees, 2002). Further applications of the concepts in signal compression, using LSPs, was briefly reviewed.

CHAPTER 8

Linear Prediction Theory for Vector Processes

8.1 INTRODUCTION

In this chapter, we consider the linear prediction problem for vector WSS processes. Let $\mathbf{x}(n)$ denote the $L \times 1$ vector sequence representing a zero-mean WSS process. Thus, $E[\mathbf{x}(n)] = 0$ and the autocorrelation is

$$\mathbf{R}(k) = E[\mathbf{x}(n)\mathbf{x}^\dagger(n\text{-}k)], \tag{8.1}$$

where $\mathbf{x}^\dagger(n)$ represents transpose conjugation, as usual. Note that $\mathbf{R}(k)$ is independent of n (owing to WSS property) and depends only on k. The autocorrelation is a matrix sequence with each element $\mathbf{R}(k)$ representing an $L \times L$ matrix. Furthermore, it is readily shown from the definition (8.1) that

$$\mathbf{R}^\dagger(-k) = \mathbf{R}(k). \tag{8.2}$$

For a vector WSS process characterized by autocorrelation $\mathbf{R}(k)$, we now formulate the linear prediction problem. Most of the developments will be along lines similar to the scalar case. But there are some differences between the scalar and vector cases that need to be emphasized. A short and sweet introduction to this topic was given in Section 9.6 of Anderson and Moore (1979). We shall go into considerably greater details in this chapter.

8.2 FORMULATION OF THE VECTOR LPC PROBLEM

The forward predictor of order N predicts the sample $\mathbf{x}(n)$ from a linear combination of past N samples:

$$\widehat{\mathbf{x}}(n) = -\mathbf{a}_{N,1}^\dagger\mathbf{x}(n-1) - \mathbf{a}_{N,2}^\dagger\mathbf{x}(n-2)\ldots - \mathbf{a}_{N,N}^\dagger\mathbf{x}(n\text{-}N). \tag{8.3}$$

The forward prediction error is therefore

$$\mathbf{e}_N^f(n) = \mathbf{x}(n) - \widehat{\mathbf{x}}(n) = \mathbf{x}(n) + \mathbf{a}_{N,1}^\dagger\mathbf{x}(n-1) + \mathbf{a}_{N,2}^\dagger\mathbf{x}(n-2)\ldots + \mathbf{a}_{N,N}^\dagger\mathbf{x}(n-N) \tag{8.4}$$

and the forward **prediction polynomial** is

$$\mathbf{A}_N(z) = \mathbf{I}_L + \mathbf{a}_{N,1}^\dagger z^{-1} + \mathbf{a}_{N,2}^\dagger z^{-2} + \ldots + \mathbf{a}_{N,N}^\dagger z^{-N}. \tag{8.5}$$

The optimal predictor has the $L \times L$ matrix coefficients $\mathbf{a}_{N,k}$ chosen such that the total mean square error in all components is minimized, that is,

$$\mathcal{E}_N^f \triangleq E[(\mathbf{e}_N^f(n))^\dagger \mathbf{e}_N^f(n)] \tag{8.6}$$

is minimized. This quantity is also equal to the trace (i.e., sum of diagonal elements) of the forward prediction *error covariance matrix*

$$\mathbf{E}_N^f = E[\mathbf{e}_N^f(n)(\mathbf{e}_N^f(n))^\dagger]. \tag{8.7}$$

Note that this is an $L \times L$ matrix. The goal therefore is to minimize

$$\mathcal{E}_N^f \triangleq \mathrm{Tr}(\mathbf{E}_N^f) \tag{8.8}$$

by choosing $\mathbf{a}_{N,k}$ appropriately. The theory of optimal predictors for the vector case is again based on the orthogonality principle:

Theorem 8.1. *Orthogonality principle (vector processes).* The Nth-order forward linear predictor for a WSS vector process $\mathbf{x}(n)$ is optimal if and only if the error $\mathbf{e}_N^f(n)$ at any time n is orthogonal to the N past observations $\mathbf{x}(n-1), \ldots, \mathbf{x}(n-N)$, that is,

$$E[\mathbf{e}_N^f(n)\mathbf{x}^\dagger(n-k)] = \mathbf{0}, \quad 1 \le k \le N, \tag{8.9}$$

where the right-hand side $\mathbf{0}$ represents the $L \times L$ zero-matrix (with L denoting the size of the column vectors $\mathbf{x}(n)$). $\diamond$

Proof. Let $\widehat{\mathbf{x}}_\perp(n)$ be the predicted value for which the error $\mathbf{e}_\perp(n)$ satisfies the orthogonality condition (8.9), and let $\widehat{\mathbf{x}}(n)$ be another predicted value with error $\mathbf{e}(n)$ (superscript f and subscript N deleted for simplicity). Now,

$$\mathbf{e}(n) = \mathbf{x}(n) - \widehat{\mathbf{x}}(n) = \mathbf{x}(n) - \widehat{\mathbf{x}}_\perp(n) + \widehat{\mathbf{x}}_\perp(n) - \widehat{\mathbf{x}}(n),$$

so that

$$\mathbf{e}(n) = \mathbf{e}_\perp(n) + \widehat{\mathbf{x}}_\perp(n) - \widehat{\mathbf{x}}(n). \tag{8.10}$$

Now, the estimates $\widehat{\mathbf{x}}_\perp(n)$ and $\widehat{\mathbf{x}}(n)$ are, by construction, linear combinations of $\mathbf{x}(n-k), 1 \le k \le N$. Because the error $\mathbf{e}_\perp(n)$, by definition, is orthogonal to these past samples, it follows that

$$\underbrace{E[\mathbf{e}(n)\mathbf{e}^\dagger(n)]}_{\text{call this } \mathbf{E}} = \underbrace{E[\mathbf{e}_\perp(n)\mathbf{e}_\perp^\dagger(n)]}_{\text{call this } \mathbf{E}_\perp} + E[(\widehat{\mathbf{x}}_\perp(n) - \widehat{\mathbf{x}}(n))(\widehat{\mathbf{x}}_\perp(n) - \widehat{\mathbf{x}}(n))^\dagger] \tag{8.11}$$

Because the second term on the right-hand side is a covariance, it is positive semidefinite. The preceding therefore shows that

$$\mathbf{E} - \mathbf{E}_\perp \geq \mathbf{0} \qquad (8.12)$$

where the notation $\mathbf{A} \geq \mathbf{0}$ means that the Hermitian matrix $\mathbf{A}$ is positive semidefinite. Taking trace on both sides, we therefore obtain

$$\mathcal{E} \geq \mathcal{E}_\perp. \qquad (8.13)$$

When does equality arise? Observe first that $\mathcal{E} - \mathcal{E}_\perp = \text{Tr}(\mathbf{E} - \mathbf{E}_\perp)$. Because $\mathbf{E} - \mathbf{E}_\perp$ has been shown to be positive semidefinite, its trace is zero only when[1] $\mathbf{E} - \mathbf{E}_\perp = \mathbf{0}$. From (8.11), we see that this is equivalent to the condition

$$E[(\widehat{\mathbf{x}}_\perp(n) - \widehat{\mathbf{x}}(n))(\widehat{\mathbf{x}}_\perp(n) - \widehat{\mathbf{x}}(n))^\dagger] = \mathbf{0},$$

which, in turn, is equivalent to $\widehat{\mathbf{x}}_\perp(n) - \widehat{\mathbf{x}}(n) = \mathbf{0}$. This proves that equality is possible if and only if $\widehat{\mathbf{x}}(n) = \widehat{\mathbf{x}}_\perp(n)$.   □

Optimal Error Covariance. The error covariance $\mathbf{E}_N^f$ for the optimal predictor can be expressed in an elegant way using the orthogonality conditions:

$$\mathbf{E}_N^f = E[\mathbf{e}_N^f(n)(\mathbf{e}_N^f(n))^\dagger] = E[\mathbf{e}_N^f(n)\left(\mathbf{x}(n) - \widehat{\mathbf{x}}(n)\right)^\dagger] = E[\mathbf{e}_N^f(n)\mathbf{x}^\dagger(n)].$$

The third equality follows by observing that the optimal $\widehat{\mathbf{x}}(n)$ is a linear combination of samples $\mathbf{x}(n - k)$, which are orthogonal to $\mathbf{e}_N^f(n)$. Using Eq. (8.4), this can be rewritten as

$$\mathbf{E}_N^f = E\left[\left(\mathbf{x}(n) + \mathbf{a}_{N,1}^\dagger\mathbf{x}(n-1) + \mathbf{a}_{N,2}^\dagger\mathbf{x}(n-2)\ldots + \mathbf{a}_{N,N}^\dagger\mathbf{x}(n-N)\right)\mathbf{x}^\dagger(n)\right]$$
$$= \mathbf{R}(0) + \mathbf{a}_{N,1}^\dagger\mathbf{R}^\dagger(1) + \mathbf{a}_{N,2}^\dagger\mathbf{R}^\dagger(2) + \ldots + \mathbf{a}_{N,N}^\dagger\mathbf{R}^\dagger(N)$$

Because the error covariance $\mathbf{E}_N^f$ is Hermitian anyway, we can rewrite this as

$$\mathbf{E}_N^f = \mathbf{R}(0) + \mathbf{R}(1)\mathbf{a}_{N,1} + \mathbf{R}(2)\mathbf{a}_{N,2} + \ldots + \mathbf{R}(N)\mathbf{a}_{N,N}. \qquad (8.14)$$

We shall find this useful when we write the augmented normal equations for optimal linear prediction.

---

[1] The trace of a matrix $\mathbf{A}$ is the sum of its eigenvalues. When $\mathbf{A}$ is positive semidefinite, all eigenvalues are nonnegative. So, zero-trace implies all eigenvalues are zero. This implies that the matrix itself is zero (because any positive semidefinite matrix can be written as $\mathbf{A} = \mathbf{U}\mathbf{\Lambda}\mathbf{U}^\dagger$, where $\mathbf{\Lambda}$ is the diagonal matrix of eigenvalues and $\mathbf{U}$ is unitary).

8.3 NORMAL EQUATIONS: VECTOR CASE

Substituting the expression (8.4) for the prediction error into the orthogonality condition (8.9), we get N matrix equations. By using the definition of autocorrelation given in Eq. (8.1), we find these equations to be

$$\mathbf{R}(k) + \mathbf{a}_{N,1}^{\dagger}\mathbf{R}(k-1) + \mathbf{a}_{N,2}^{\dagger}\mathbf{R}(k-2) + \ldots + \mathbf{a}_{N,N}^{\dagger}\mathbf{R}(k-N) = \mathbf{0}, \qquad (8.15)$$

for $1 \leq k \leq N$. Here the $\mathbf{0}$ on the right is an $L \times L$ matrix of zeros. Solving these equations, we can obtain the N optimal predictor coefficient matrices $\mathbf{a}_{N,k}$. For convenience, we shall rewrite the equations in the form

$$\mathbf{R}^{\dagger}(k-1)\mathbf{a}_{N,1} + \mathbf{R}^{\dagger}(k-2)\mathbf{a}_{N,2} + \ldots + \mathbf{R}^{\dagger}(k-N)\mathbf{a}_{N,N} = -\mathbf{R}^{\dagger}(k) \qquad (8.16)$$

for $1 \leq k \leq N$. By using the fact that $\mathbf{R}^{\dagger}(-k) = \mathbf{R}(k)$, these equations can be written in the following elegant form:

$$\begin{bmatrix} \mathbf{R}(0) & \mathbf{R}(1) & \ldots \mathbf{R}(N-1) \\ \mathbf{R}(-1) & \mathbf{R}(0) & \ldots \mathbf{R}(N-2) \\ \vdots & \vdots & \ddots & \vdots \\ \mathbf{R}(-N+1) & \mathbf{R}(-N+2) & \ldots & \mathbf{R}(0) \end{bmatrix} \begin{bmatrix} \mathbf{a}_{N,1} \\ \mathbf{a}_{N,2} \\ \vdots \\ \mathbf{a}_{N,N} \end{bmatrix} = - \begin{bmatrix} \mathbf{R}(-1) \\ \mathbf{R}(-2) \\ \vdots \\ \mathbf{R}(-N) \end{bmatrix}. \qquad (8.17)$$

These represent the *normal equations* for the vector LPC problem. Note that these equations reduce to the normal equations for scalar processes given in Eq. (2.8) (remembering $\mathbf{R}^{\dagger}(-k) = \mathbf{R}(k)$). If we move the vector on the right to the left-hand side and then prefix Eq. (8.14) at the top, we get the following *augmented normal equations* for the optimal vector LPC problem:

$$\begin{bmatrix} \mathbf{R}(0) & \mathbf{R}(1) & \ldots & \mathbf{R}(N) \\ \mathbf{R}(-1) & \mathbf{R}(0) & \ldots & \mathbf{R}(N-1) \\ \vdots & \vdots & \ddots & \vdots \\ \mathbf{R}(-N) & \mathbf{R}(-N+1) & \ldots & \mathbf{R}(0) \end{bmatrix} \begin{bmatrix} \mathbf{I}_{L} \\ \mathbf{a}_{N,1} \\ \vdots \\ \mathbf{a}_{N,N} \end{bmatrix} = \begin{bmatrix} \mathbf{E}_{N}^{f} \\ 0 \\ \vdots \\ 0 \end{bmatrix}. \qquad (8.18)$$

We will soon return to this equation and solve it. But first, we have to introduce the idea of backward prediction.

8.4 BACKWARD PREDICTION

The formulation of the backward predictor problem for the vector case allows us to derive a recursion similar to Levinson's recursion in the scalar case (Chapter 3). This also results in elegant lattice structures. The Nth-order backward linear predictor predicts the $L \times 1$ vector $\mathbf{x}(n - N - 1)$ as follows:

$$\widehat{\mathbf{x}}(n - N - 1) = -\mathbf{b}_{N,1}^{\dagger}\mathbf{x}(n - 1) - \mathbf{b}_{N,2}^{\dagger}\mathbf{x}(n - 2) - \ldots - \mathbf{b}_{N,N}^{\dagger}\mathbf{x}(n - N),$$

where $\mathbf{b}_{N,k}$ are $L \times L$ matrices. The backward prediction error is

$$\mathbf{e}_N^{b}(n) = \mathbf{x}(n - N - 1) - \widehat{\mathbf{x}}(n - N - 1),$$

so that

$$\mathbf{e}_N^{b}(n) = \mathbf{x}(n - N - 1) + \mathbf{b}_{N,1}^{\dagger}\mathbf{x}(n - 1) + \mathbf{b}_{N,2}^{\dagger}\mathbf{x}(n - 2) + \ldots + \mathbf{b}_{N,N}^{\dagger}\mathbf{x}(n - N).$$

So the backward prediction polynomial takes the form

$$\mathbf{B}_N(z) = \mathbf{b}_{N,1}^{\dagger}z^{-1} + \mathbf{b}_{N,2}^{\dagger}z^{-2} \ldots + \mathbf{b}_{N,N}^{\dagger}z^{-N} + z^{-(N+1)}\mathbf{I}. \tag{8.19}$$

The optimal coefficients $\mathbf{b}_{N,k}$ are again found by imposing the orthogonality condition (8.9), namely,

$$E[\mathbf{e}_N^{b}(n)\mathbf{x}^{\dagger}(n - k)] = \mathbf{0}, \quad 1 \le k \le N$$

This results in the N equations

$$\mathbf{b}_{N,1}^{\dagger}\mathbf{R}(k - 1) + \mathbf{b}_{N,2}^{\dagger}\mathbf{R}(k - 2) + \ldots \mathbf{b}_{N,N}^{\dagger}\mathbf{R}(k - N) + \mathbf{R}(k - N - 1) = \mathbf{0} \tag{8.20}$$

for $1 \le k \le N$. The error covariance for the optimal backward predictor is

$$\begin{aligned}\mathbf{E}_N^{b} &= E[\mathbf{e}_N^{b}(n)(\mathbf{e}_N^{b}(n))^{\dagger}] = E[\mathbf{e}_N^{b}(n)\,(\mathbf{x}(n - N - 1) - \widehat{\mathbf{x}}(n - N - 1))^{\dagger}] \\ &= E[\mathbf{e}_N^{b}(n)(\mathbf{x}^{\dagger}(n - N - 1)].\end{aligned}$$

The third equality follows by observing that the optimal $\widehat{\mathbf{x}}(n - N - 1)$ is a linear combination of samples $\mathbf{x}(n - k)$ that are orthogonal to $\mathbf{e}_N^{b}(n)$. So $\mathbf{E}_N^{b}$ can be rewritten as

$$\mathbf{E}_N^{b} = E\left[\left(\mathbf{x}(n - N - 1) + \mathbf{b}_{N,1}^{\dagger}\mathbf{x}(n - 1) + \ldots + \mathbf{b}_{N,N}^{\dagger}\mathbf{x}(n - N)\right)\mathbf{x}^{\dagger}(n - N - 1)\right]$$

Using Eq. (8.1) and the facts that $\mathbf{E}_N^{b}$ is Hermitian and $\mathbf{R}^{\dagger}(-k) = \mathbf{R}(k)$, this can further be rewritten as

$$\mathbf{E}_N^{b} = \mathbf{R}(-N)\mathbf{b}_{N,1} + \ldots + \mathbf{R}(-1)\mathbf{b}_{N,N} + \mathbf{R}(0). \tag{8.21}$$

Combining this with the N equations in Eq. (8.20) and rearranging slightly, we obtain the following *augmented normal equations* for the backward vector predictor:

$$
\begin{bmatrix}
\mathbf{R}(0) & \mathbf{R}(1) & \ldots & \mathbf{R}(N) \\
\mathbf{R}(-1) & \mathbf{R}(0) & \ldots & \mathbf{R}(N-1) \\
\vdots & \vdots & \ddots & \vdots \\
\mathbf{R}(-N) & \mathbf{R}(-N+1) & \ldots & \mathbf{R}(0)
\end{bmatrix}
\begin{bmatrix}
\mathbf{b}_{N,1} \\
\vdots \\
\mathbf{b}_{N,N} \\
\mathbf{I}_L
\end{bmatrix}
=
\begin{bmatrix}
\mathbf{0} \\
\mathbf{0} \\
\vdots \\
\mathbf{E}_N^b
\end{bmatrix}
\tag{8.22}
$$

8.5   LEVINSON'S RECURSION: VECTOR CASE

We now combine the normal equations for the forward and backward predictors to obtain Levinson's recursion for solving the optimal predictors recursively. For this, we start from Eq. (8.18) and add an extra column and row[2] to obtain

$$
\underbrace{
\begin{bmatrix}
\mathbf{R}(0) & \mathbf{R}(1) & \ldots & \mathbf{R}(N) & \mathbf{R}(N+1) \\
\mathbf{R}(-1) & \mathbf{R}(0) & \ldots & \mathbf{R}(N-1) & \mathbf{R}(N) \\
\vdots & \vdots & \ddots & \vdots & \\
\mathbf{R}(-N) & \mathbf{R}(-N+1) & \ldots & \mathbf{R}(0) & \mathbf{R}(1) \\
\mathbf{R}(-N-1) & \mathbf{R}(-N) & \ldots & \mathbf{R}(-1) & \mathbf{R}(0)
\end{bmatrix}
}_{\text{call this } \mathbf{R}_{N+2}}
\begin{bmatrix}
\mathbf{I}_L \\
\mathbf{a}_{N,1} \\
\vdots \\
\mathbf{a}_{N,N} \\
\mathbf{0}
\end{bmatrix}
=
\begin{bmatrix}
\mathbf{E}_N^f \\
\mathbf{0} \\
\vdots \\
\mathbf{0} \\
\mathbf{F}_N^f
\end{bmatrix},
\tag{8.23}
$$

where $\mathbf{F}_N^f$ is an $L \times L$ matrix, possibly nonzero, which comes from the extra row added at the bottom. Similarly, starting from Eq. (8.22), we add an extra row and column to obtain

$$
\underbrace{
\begin{bmatrix}
\mathbf{R}(0) & \mathbf{R}(1) & \ldots & \mathbf{R}(N) & \mathbf{R}(N+1) \\
\mathbf{R}(-1) & \mathbf{R}(0) & \ldots & \mathbf{R}(N-1) & \mathbf{R}(N) \\
\vdots & \vdots & \ddots & \vdots & \\
\mathbf{R}(-N) & \mathbf{R}(-N+1) & \ldots & \mathbf{R}(0) & \mathbf{R}(1) \\
\mathbf{R}(-N-1) & \mathbf{R}(-N) & \ldots & \mathbf{R}(-1) & \mathbf{R}(0)
\end{bmatrix}
}_{\text{this is } \mathbf{R}_{N+2} \text{ too!}}
\begin{bmatrix}
\mathbf{0} \\
\mathbf{b}_{N,1} \\
\vdots \\
\mathbf{b}_{N,N} \\
\mathbf{I}_L
\end{bmatrix}
=
\begin{bmatrix}
\mathbf{F}_N^b \\
\mathbf{0} \\
\vdots \\
\mathbf{0} \\
\mathbf{E}_N^b
\end{bmatrix}
\tag{8.24}
$$

where $\mathbf{F}_N^b$ comes from the extra row added at the top. We now postmultiply Eq. (8.24) by an $L \times L$ matrix $(\mathbf{K}_{N+1}^f)^\dagger$ and add it to Eq. (8.23) with the hope that the last entry $\mathbf{F}_N^f$ in the right-hand side is

---

[2]Actually L extra columns and rows because each entry is itself an $L \times L$ matrix.

replaced with zero. This will then lead to the augmented normal equation for the $(N+1)$th-order forward predictor. We have

$$\mathbf{R}_{N+2}\left(\begin{bmatrix} \mathbf{I}_L \\ \mathbf{a}_{N,1} \\ \vdots \\ \mathbf{a}_{N,N} \\ \mathbf{0} \end{bmatrix} + \begin{bmatrix} \mathbf{0} \\ \mathbf{b}_{N,1} \\ \vdots \\ \mathbf{b}_{N,N} \\ \mathbf{I}_L \end{bmatrix}(\mathbf{K}^f_{N+1})^\dagger\right) = \begin{bmatrix} \mathbf{E}^f_N \\ \mathbf{0} \\ \vdots \\ \mathbf{0} \\ \mathbf{F}^f_N \end{bmatrix} + \begin{bmatrix} \mathbf{F}^b_N \\ \mathbf{0} \\ \vdots \\ \mathbf{0} \\ \mathbf{E}^b_N \end{bmatrix}(\mathbf{K}^f_{N+1})^\dagger \qquad (8.25)$$

Clearly, the choice of $\mathbf{K}^f_{N+1}$, which achieves the aforementioned purpose should be such that

$$\mathbf{F}^f_N + \mathbf{E}^b_N(\mathbf{K}^f_{N+1})^\dagger = \mathbf{0}.$$

At this point, we make the assumption that the $L \times L$ error covariance $\mathbf{E}^b_N$, and for future purposes, $\mathbf{E}^f_N$, are nonsingular. The solution $\mathbf{K}^f_{N+1}$ is therefore given by

$$\mathbf{K}^f_{N+1} = -[\mathbf{F}^f_N]^\dagger[\mathbf{E}^b_N]^{-1}. \qquad (8.26)$$

We can similarly postmultiply Eq. (8.23) by an $L \times L$ matrix $\mathbf{K}^b_{N+1}$ and add it to Eq. (8.24):

$$\mathbf{R}_{N+2}\left(\begin{bmatrix} \mathbf{I}_L \\ \mathbf{a}_{N,1} \\ \vdots \\ \mathbf{a}_{N,N} \\ \mathbf{0} \end{bmatrix}\mathbf{K}^b_{N+1} + \begin{bmatrix} \mathbf{0} \\ \mathbf{b}_{N,1} \\ \vdots \\ \mathbf{b}_{N,N} \\ \mathbf{I}_L \end{bmatrix}\right) = \begin{bmatrix} \mathbf{E}^f_N \\ \mathbf{0} \\ \vdots \\ \mathbf{0} \\ \mathbf{F}^f_N \end{bmatrix}\mathbf{K}^b_{N+1} + \begin{bmatrix} \mathbf{F}^b_N \\ \mathbf{0} \\ \vdots \\ \mathbf{0} \\ \mathbf{E}^b_N \end{bmatrix}. \qquad (8.27)$$

To create the augmented normal equation for the $(N+1)$-order backward predictor, we have to reduce this to the form (8.22). For this, $\mathbf{K}^b_{N+1}$ should be chosen as

$$[\mathbf{E}^f_N]\mathbf{K}^b_{N+1} + \mathbf{F}^b_N = \mathbf{0},$$

that is,

$$\mathbf{K}^b_{N+1} = -[\mathbf{E}^f_N]^{-1}\mathbf{F}^b_N. \qquad (8.28)$$

The matrices $\mathbf{K}_{N+1}^{f}$ and $\mathbf{K}_{N+1}^{b}$ are analogous to the parcor coefficients k_m used in scalar linear prediction. With $\mathbf{K}_{N+1}^{f}$ and $\mathbf{K}_{N+1}^{b}$ chosen as above, Eqs. (8.25) and (8.27) reduce to the form

$$
\mathbf{R}_{N+2}\left(\begin{bmatrix} \mathbf{I}_L \\ \mathbf{a}_{N,1} \\ \vdots \\ \mathbf{a}_{N,N} \\ \mathbf{0} \end{bmatrix} + \begin{bmatrix} \mathbf{0} \\ \mathbf{b}_{N,1} \\ \vdots \\ \mathbf{b}_{N,N} \\ \mathbf{I}_L \end{bmatrix} (\mathbf{K}_{N+1}^{f})^{\dagger}\right) = \begin{bmatrix} \mathbf{E}_{N+1}^{f} \\ \mathbf{0} \\ \vdots \\ \mathbf{0} \\ \mathbf{0} \end{bmatrix} \qquad (8.29)
$$

and

$$
\mathbf{R}_{N+2}\left(\begin{bmatrix} \mathbf{I}_L \\ \mathbf{a}_{N,1} \\ \vdots \\ \mathbf{a}_{N,N} \\ \mathbf{0} \end{bmatrix} \mathbf{K}_{N+1}^{b} + \begin{bmatrix} \mathbf{0} \\ \mathbf{b}_{N,1} \\ \vdots \\ \mathbf{b}_{N,N} \\ \mathbf{I}_L \end{bmatrix}\right) = \begin{bmatrix} \mathbf{0} \\ \mathbf{0} \\ \vdots \\ \mathbf{0} \\ \mathbf{E}_{N+1}^{b} \end{bmatrix}, \qquad (8.30)
$$

respectively. These are the augmented normal equations for the $(N+1)$th-order optimal forward and backward vector predictors, respectively. Thus, the $(N+1)$th-order forward predictor coefficients are obtained from

$$
\begin{bmatrix} \mathbf{I}_L \\ \mathbf{a}_{N+1,1} \\ \vdots \\ \mathbf{a}_{N+1,N} \\ \mathbf{a}_{N+1,N+1} \end{bmatrix} = \begin{bmatrix} \mathbf{I}_L \\ \mathbf{a}_{N,1} \\ \vdots \\ \mathbf{a}_{N,N} \\ \mathbf{0} \end{bmatrix} + \begin{bmatrix} \mathbf{0} \\ \mathbf{b}_{N,1} \\ \vdots \\ \mathbf{b}_{N,N} \\ \mathbf{I}_L \end{bmatrix} (\mathbf{K}_{N+1}^{f})^{\dagger}.
$$

Similarly, the $(N+1)$th-order backward predictor coefficients are obtained from

$$
\begin{bmatrix} \mathbf{b}_{N+1,1} \\ \mathbf{b}_{N+1,2} \\ \vdots \\ \mathbf{b}_{N+1,N+1} \\ \mathbf{I}_L \end{bmatrix} = \begin{bmatrix} \mathbf{I}_L \\ \mathbf{a}_{N,1} \\ \vdots \\ \mathbf{a}_{N,N} \\ \mathbf{0} \end{bmatrix} \mathbf{K}_{N+1}^{b} + \begin{bmatrix} \mathbf{0} \\ \mathbf{b}_{N,1} \\ \vdots \\ \mathbf{b}_{N,N} \\ \mathbf{I}_L \end{bmatrix}.
$$

The forward and backward prediction polynomials defined in Eqs. (8.5) and (8.19) are therefore given by

$$
\mathbf{A}_{N+1}(z) = \mathbf{A}_N(z) + \mathbf{K}_{N+1}^{f} \mathbf{B}_N(z) \qquad (8.31)
$$

and

$$\mathbf{B}_{N+1}(z) = z^{-1}[(\mathbf{K}^{\mathrm{b}}_{N+1})^{\dagger}\mathbf{A}_N(z) + \mathbf{B}_N(z)]. \tag{8.32}$$

The extra z^{-1} arises because the 0th coefficient in the backward predictor, by definition, is zero (see Eq. (8.19)). By comparing Eqs. (8.25) and (8.29) we see that the optimal prediction error covariance matrix is updated as follows:

$$\mathbf{E}^f_{N+1} = \mathbf{E}^f_N + \mathbf{F}^{\mathrm{b}}_N(\mathbf{K}^f_{N+1})^{\dagger}. \tag{8.33}$$

Similarly, from (8.25) and (8.29), we obtain

$$\mathbf{E}^{\mathrm{b}}_{N+1} = \mathbf{E}^{\mathrm{b}}_N + \mathbf{F}^f_N\mathbf{K}^{\mathrm{b}}_{N+1}. \tag{8.34}$$

By substituting from Eq. (8.26) and (8.28) and using the fact that the error covariances are Hermitian anyway, we obtain the error covariance update equations

$$\mathbf{E}^f_{N+1} = \left(\mathbf{I}_L - \mathbf{K}^f_{N+1}(\mathbf{K}^{\mathrm{b}}_{N+1})^{\dagger}\right)\mathbf{E}^f_N \tag{8.35}$$

and

$$\mathbf{E}^{\mathrm{b}}_{N+1} = \left(\mathbf{I}_L - (\mathbf{K}^{\mathrm{b}}_{N+1})^{\dagger}\mathbf{K}^f_{N+1}\right)\mathbf{E}^{\mathrm{b}}_N \tag{8.36}$$

Summary of Levinson's Recursion. We now summarize the key equations of Levinson's recursion. The subscript m is used instead of N for future convenience. Let $\mathbf{R}(k)$ be the $L \times L$ autocorrelation matrix sequence of a WSS process $\mathbf{x}(n)$. Then, the optimal linear predictor polynomials $\mathbf{A}_N(z)$ and $\mathbf{B}_N(z)$ and the error covariances $\mathbf{E}^f_N$ and $\mathbf{E}^{\mathrm{b}}_N$ can be calculated recursively as follows: first, initialize the recursion according to

$$\mathbf{A}_0(z) = \mathbf{I}, \quad \mathbf{B}_0(z) = z^{-1}\mathbf{I}, \quad \text{and} \quad \mathbf{E}^f_0 = \mathbf{E}^{\mathrm{b}}_0 = \mathbf{R}(0). \tag{8.37(a)}$$

Given $\mathbf{A}_m(z)$, $\mathbf{B}_m(z)$, $\mathbf{E}^f_m$, and $\mathbf{E}^{\mathrm{b}}_m$ (all $L \times L$ matrices), proceed as follows:

1. Calculate the $L \times L$ matrices

$$\mathbf{F}^f_m = \mathbf{R}(-m-1) + \mathbf{R}(-m)\mathbf{a}_{m,1} + \ldots + \mathbf{R}(-1)\mathbf{a}_{m,m} \tag{8.37(b)}$$

$$\mathbf{F}^{\mathrm{b}}_m = \mathbf{R}(1)\mathbf{b}_{m,1} + \ldots + \mathbf{R}(m)\mathbf{b}_{m,m} + \mathbf{R}(m+1) \tag{8.37(c)}$$

2. Calculate the parcor coefficients ($L \times L$ matrices)

$$\mathbf{K}^f_{m+1} = -[\mathbf{F}^f_m]^{\dagger}[\mathbf{E}^{\mathrm{b}}_m]^{-1} \tag{8.37(d)}$$

$$\mathbf{K}^{\mathrm{b}}_{m+1} = -[\mathbf{E}^f_m]^{-1}\mathbf{F}^{\mathrm{b}}_m \tag{8.37(e)}$$

3. Update the prediction polynomials according to

$$\mathbf{A}_{m+1}(z) = \mathbf{A}_m(z) + \mathbf{K}_{m+1}^f \mathbf{B}_m(z) \qquad (8.37(f))$$

$$\mathbf{B}_{m+1}(z) = z^{-1}[(\mathbf{K}_{m+1}^b)^\dagger \mathbf{A}_m(z) + \mathbf{B}_m(z)]. \qquad (8.37(g))$$

4. Update the error covariances according to

$$\mathbf{E}_{m+1}^f = \left(\mathbf{I}_L - \mathbf{K}_{m+1}^f(\mathbf{K}_{m+1}^b)^\dagger\right)\mathbf{E}_m^f \qquad (8.37(h))$$

$$\mathbf{E}_{m+1}^b = \left(\mathbf{I}_L - (\mathbf{K}_{m+1}^b)^\dagger\mathbf{K}_{m+1}^f\right)\mathbf{E}_m^b. \qquad (8.37(i))$$

This recursion can be repeated to obtain the optimal predictors for any order N.

8.6   PROPERTIES DERIVED FROM LEVINSON'S RECURSION

We now derive some properties of the matrices that are involved in Levinson's recursion. This gives additional insight and places the mathematical equations in proper context.

8.6.1   Properties of Matrices $\mathbf{F}_m^f$ and $\mathbf{F}_m^b$

Eq. (8.37(b)) can be rewritten as

$$\begin{aligned}
(\mathbf{F}_m^f)^\dagger &= \mathbf{R}(m+1) + \mathbf{a}_{m,1}^\dagger\mathbf{R}(m) + \ldots + \mathbf{a}_{m,m}^\dagger\mathbf{R}(1) \\
&= E\left[\left(\mathbf{x}(n) + \mathbf{a}_{m,1}^\dagger\mathbf{x}(n-1) + \ldots + \mathbf{a}_{m,m}^\dagger\mathbf{x}(n-m)\right)\mathbf{x}^\dagger(n-m-1)\right] \\
&= E[\mathbf{e}_m^f(n)\mathbf{x}^\dagger(n-m-1)]
\end{aligned}$$

We know the error $\mathbf{e}_m^f(n)$ is orthogonal to the samples $\mathbf{x}(n-1), \ldots, \mathbf{x}(n-m)$, but not necessarily to $\mathbf{x}(n-m-1)$. This "residual correlation" between $\mathbf{e}_m^f(n)$ and $\mathbf{x}(n-m-1)$ is precisely what is captured by the matrix $\mathbf{F}_m^f$. The coefficient $\mathbf{K}_{m+1}^f$ in Eq. (8.37(d)) is therefore called the *partial correlation* or *parcor* coefficient, as in the scalar case. From Eq. (8.37(d)), we see that if this "residual correlation" is zero, then $\mathbf{K}_{m+1}^f = 0$, and as a consequence, there is no progress in the recursion, that is,

$$\mathbf{A}_{m+1}(z) = \mathbf{A}_m(z) \quad \text{and} \quad \mathbf{E}_{m+1}^f = \mathbf{E}_m^f.$$

How about $\mathbf{F}_m^b$? Does it have a similar significance? Observe from (8.37(c)) that

$$\begin{aligned}
(\mathbf{F}_m^b)^\dagger &= \mathbf{b}_{m,1}^\dagger\mathbf{R}(-1) + \ldots + \mathbf{b}_{m,m}^\dagger\mathbf{R}(-m) + \mathbf{R}(-m-1) \\
&= E\left[\left(\mathbf{b}_{m,1}^\dagger\mathbf{x}(n-1) + \ldots + \mathbf{b}_{m,m}^\dagger\mathbf{x}(n-m) + \mathbf{x}(n-m-1)\right)\mathbf{x}^\dagger(n)\right] \\
&= E[\mathbf{e}_m^b(n)\mathbf{x}^\dagger(n)]
\end{aligned}$$

The backward error $\mathbf{e}_m^b(n)$ is the error in the estimation of $\mathbf{x}(n - m - 1)$ based on the samples $\mathbf{x}(n - m), \ldots, \mathbf{x}(n - 1)$. Thus, $\mathbf{e}_m^b(n)$ is orthogonal to $\mathbf{x}(n - m), \ldots, \mathbf{x}(n - 1)$ but not necessarily to the sample $\mathbf{x}(n)$. This residual correlation is captured by $\mathbf{F}_m^b$. If this is zero, then $\mathbf{K}_{m+1}^b$ is zero as seen from Eq. (8.37(e)). That is, there is no progress in prediction as we go from stage m to $m + 1$.

Relation Between $\mathbf{F}_m^f$ and $\mathbf{F}_m^b$. We have shown that

$$(\mathbf{F}_m^f)^\dagger = E[\mathbf{e}_m^f(n)\mathbf{x}^\dagger(n - m - 1)] \tag{8.38}$$

and

$$(\mathbf{F}_m^b)^\dagger = E[\mathbf{e}_m^b(n)\mathbf{x}^\dagger(n)]. \tag{8.39}$$

For the scalar case, the parameter α_m played the role of the matrices $\mathbf{F}_m^f$ and $\mathbf{F}_m^b$. But for the vector case, we have the two matrices $\mathbf{F}_m^f$ and $\mathbf{F}_m^b$, which appear to be unrelated. This gives the first impression that there is a lack of symmetry between the forward and backward predictors. However, there is a simple relation between $\mathbf{F}_m^f$ and $\mathbf{F}_m^b$ credited to Burg (Kailath, 1974); namely,

$$\mathbf{F}_m^b = (\mathbf{F}_m^f)^\dagger. \tag{8.40}$$

This is readily proved as follows:

Proof of Eq. (8.40). We can rewrite Eq. (8.38) as

$$(\mathbf{F}_m^f)^\dagger = E\left[\mathbf{e}_m^f(n)\left(\mathbf{x}(n - m - 1) + \mathbf{b}_{m,1}^\dagger\mathbf{x}(n - 1) + \ldots + \mathbf{b}_{m,m}^\dagger\mathbf{x}(n - m)\right)^\dagger\right]$$
$$= E[\mathbf{e}_m^f(n)(\mathbf{e}_m^b(n))^\dagger]$$

because $E[\mathbf{e}_m^f(n)\mathbf{x}(n - k)] = \mathbf{0}$ for $1 \leq k \leq m$ anyway. Similarly, from Eq. (8.39), we get

$$(\mathbf{F}_m^b)^\dagger = E\left[\mathbf{e}_m^b(n)\left[\mathbf{x}(n) + \mathbf{a}_{m,1}^\dagger\mathbf{x}(n - 1) + \ldots + \mathbf{a}_{m,m}^\dagger\mathbf{x}(n - m)\right]^\dagger\right]$$
$$= E[\mathbf{e}_m^b(n)(\mathbf{e}_m^f(n))^\dagger].$$

So we have shown that

$$\mathbf{F}_m^f = E[\mathbf{e}_m^b(n)(\mathbf{e}_m^f(n))^\dagger] \quad \text{and} \quad \mathbf{F}_m^b = E[\mathbf{e}_m^f(n)(\mathbf{e}_m^b(n))^\dagger], \tag{8.41}$$

which proves (8.40). $\qquad\qquad\qquad\qquad\qquad\qquad\qquad\qquad\qquad\qquad\qquad\square$

Relation Between Error Covariance Matrices. From the definitions of $\mathbf{K}_{m+1}^f$ and $\mathbf{K}_{m+1}^b$ given in Eqs. (8.37(d)) and (8.37(e)), we have

$$[\mathbf{F}_m^f]^\dagger = -\mathbf{K}_{m+1}^f\mathbf{E}_m^b \quad \text{and} \quad \mathbf{F}_m^b = -\mathbf{E}_m^f\mathbf{K}_{m+1}^b.$$

In view of the symmetry relation (8.40), it then follows that

$$\mathbf{K}_{m+1}^{f}\mathbf{E}_{m}^{b} = \mathbf{E}_{m}^{f}\mathbf{K}_{m+1}^{b}. \tag{8.42}$$

8.6.2 Monotone Properties of the Error Covariance Matrices

We now show that the error covariances of the optimal linear predictor satisfy the following:

$$\mathbf{E}_{m}^{f} \geq \mathbf{E}_{m+1}^{f} \quad \text{and} \quad \mathbf{E}_{m}^{b} \geq \mathbf{E}_{m+1}^{b}, \tag{8.43}$$

that is, the differences $\mathbf{E}_{m}^{f} - \mathbf{E}_{m+1}^{f}$ and $\mathbf{E}_{m}^{b} - \mathbf{E}_{m+1}^{b}$ are positive semidefinite. Because the total mean square error $\mathcal{E}_{m}^{f}$ is the trace of $\mathbf{E}_{m}^{f}$, Eq. (8.43) means in particular that

$$\mathcal{E}_{m}^{f} \geq \mathcal{E}_{m+1}^{f} \quad \text{and} \quad \mathcal{E}_{m}^{b} \geq \mathcal{E}_{m+1}^{b} \tag{8.44}$$

Proof of (8.43). Substituting (8.37(d)) into Eq. (8.33), we see that $\mathbf{E}_{m+1}^{f} = \mathbf{E}_{m}^{f} - \mathbf{F}_{m}^{b}(\mathbf{E}_{m}^{b})^{-1}\mathbf{F}_{m}^{f}$. In view of the symmetry relation (8.40), it then follows that

$$\mathbf{E}_{m+1}^{f} = \mathbf{E}_{m}^{f} - (\mathbf{F}_{m}^{f})^{\dagger}(\mathbf{E}_{m}^{b})^{-1}\mathbf{F}_{m}^{f}. \tag{8.45}$$

Because $(\mathbf{E}_{m}^{b})^{-1}$ is Hermitian and positive definite, it follows that the second term on the right-hand side is Hermitian and positive semidefinite. This proves $\mathbf{E}_{m}^{f} \geq \mathbf{E}_{m+1}^{f}$. Similarly, substituting Eq. (8.37(e)) into Eq. (8.34), we get $\mathbf{E}_{m+1}^{b} = \mathbf{E}_{m}^{b} - \mathbf{F}_{m}^{f}(\mathbf{E}_{m}^{f})^{-1}\mathbf{F}_{m}^{b}$, from which it follows that

$$\mathbf{E}_{m+1}^{b} = \mathbf{E}_{m}^{b} - \mathbf{F}_{m}^{f}(\mathbf{E}_{m}^{f})^{-1}(\mathbf{F}_{m}^{f})^{\dagger}. \tag{8.46}$$

This shows that $\mathbf{E}_{m}^{b} \geq \mathbf{E}_{m+1}^{b}$ and the proof of Eq. (8.43) is complete. □

8.6.3 Summary of Properties Relating to Levinson's Recursion

Here is a summary of what we have shown so far:

1. $\mathbf{F}_{m}^{f}$ and $\mathbf{F}_{m}^{b}$ have the following interpretations:

$$(\mathbf{F}_{m}^{f})^{\dagger} = E[\mathbf{e}_{m}^{f}(n)\mathbf{x}^{\dagger}(n-m-1)], \quad (\mathbf{F}_{m}^{b})^{\dagger} = E[\mathbf{e}_{m}^{b}(n)\mathbf{x}^{\dagger}(n)]. \tag{8.47(a)}$$

2. We can also express these as

$$\mathbf{F}_{m}^{f} = E[\mathbf{e}_{m}^{b}(n)(\mathbf{e}_{m}^{f}(n))^{\dagger}] \quad \text{and} \quad \mathbf{F}_{m}^{b} = E[\mathbf{e}_{m}^{f}(n)(\mathbf{e}_{m}^{b}(n))^{\dagger}], \tag{8.47(b)}$$

so that

$$\mathbf{F}_{m}^{b} = (\mathbf{F}_{m}^{f})^{\dagger}. \tag{8.47(c)}$$

3. The parcor coefficients are related as

$$\mathbf{K}_{m+1}^{f}\mathbf{E}_{m}^{b} = \mathbf{E}_{m}^{f}\mathbf{K}_{m+1}^{b}. \qquad (8.47(d))$$

4. The error covariances matrices satisfy

$$\mathbf{E}_{m}^{f} \geq \mathbf{E}_{m+1}^{f}, \quad \mathbf{E}_{m}^{b} \geq \mathbf{E}_{m+1}^{b}, \qquad (8.47(e))$$

   where $\mathbf{A} \geq \mathbf{B}$ means that $\mathbf{A} - \mathbf{B}$ is positive semidefinite.

5. Traces of the error covariances satisfy $\mathcal{E}_{m}^{f} \geq \mathcal{E}_{m+1}^{f}$ and $\mathcal{E}_{m}^{b} \geq \mathcal{E}_{m+1}^{b}$.

6. For completeness, we list one more relation that will only be proved in Section 8.8.1. The error covariances for successive prediction orders are related as

$$(\mathbf{E}_{m+1}^{f})^{1/2} = (\mathbf{E}_{m}^{f})^{1/2}\left(\mathbf{I}_{L} - \mathbf{P}_{m+1}\mathbf{P}_{m+1}^{\dagger}\right)^{1/2} \qquad (8.47(f))$$

$$(\mathbf{E}_{m+1}^{b})^{1/2} = (\mathbf{E}_{m}^{b})^{1/2}\left(\mathbf{I}_{L} - \mathbf{P}_{m+1}^{\dagger}\mathbf{P}_{m+1}\right)^{1/2}, \qquad (8.47(g))$$

where

$$\mathbf{P}_{m+1} \triangleq (\mathbf{E}_{m}^{f})^{-1/2}\mathbf{K}_{m+1}^{f}(\mathbf{E}_{m}^{b})^{1/2} = (\mathbf{E}_{m}^{f})^{\dagger/2}\mathbf{K}_{m+1}^{b}(\mathbf{E}_{m}^{b})^{-\dagger/2} \qquad (8.47(h))$$

and the matrices $\mathbf{P}_{m}$ satisfy

$$\mathbf{P}_{m}\mathbf{P}_{m}^{\dagger} < \mathbf{I}, \quad \mathbf{P}_{m}^{\dagger}\mathbf{P}_{m} < \mathbf{I}. \qquad (8.47(i))$$

8.7   TRANSFER MATRIX FUNCTIONS IN VECTOR LPC

The forward prediction polynomial (8.5) is a multi-input multi-output (MIMO) FIR filter, which produces the output $\mathbf{e}_{N}^{f}(n)$ in response to the input $\mathbf{x}(n)$. Similarly backward prediction polynomial (8.19) produces $\mathbf{e}_{N}^{b}(n)$ in response to the input $\mathbf{x}(n)$. These are schematically shown in Fig. 8.1(a). It then follows that the backward error $\mathbf{e}_{N}^{b}(n)$ can be generated from the forward error $\mathbf{e}_{N}^{f}(n)$ as shown in part (b) of the figure. We therefore see that the transfer function from $\mathbf{e}_{N}^{f}(n)$ to $\mathbf{e}_{N}^{b}(n)$ is given by

$$z^{-1}\mathbf{H}_{N}(z) \triangleq \mathbf{B}_{N}(z)\mathbf{A}_{N}^{-1}(z) \qquad (8.48)$$

A modification of this is shown in part (c) where the quantities $\mathbf{g}_{N}^{f}(n)$ and $\mathbf{g}_{N}^{b}(n)$ are given by

$$\mathbf{g}_{N}^{f}(n) = (\mathbf{E}_{N}^{f})^{-1/2}\mathbf{e}_{N}^{f}(n), \quad \mathbf{g}_{N}^{b}(n) = (\mathbf{E}_{N}^{b})^{-1/2}\mathbf{e}_{N}^{b}(n) \qquad (8.49)$$

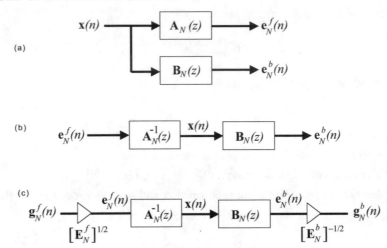

FIGURE 8.1: (a) Generation of the forward and backward errors using the prediction polynomials $\mathbf{A}_N(z)$ and $\mathbf{B}_N(z)$, (b) generation of the backward error $\mathbf{e}_N^{\mathrm{b}}(n)$ from the forward error $\mathbf{e}_N^f(n)$, and (c) generation of the normalized backward error $\mathbf{g}_N^{\mathrm{b}}(n)$ from the normalized forward error $\mathbf{g}_N^f(n)$.

These are called *normalized errors* because their covariance matrices are equal to $\mathbf{I}$. The transfer function from $\mathbf{g}_N^f(n)$ to $\mathbf{g}_N^{\mathrm{b}}(n)$ is given by

$$z^{-1}\mathbf{G}_N(z) \triangleq [\mathbf{E}_N^{\mathrm{b}}]^{-1/2}\mathbf{B}_N(z)\mathbf{A}_N^{-1}(z)[\mathbf{E}_N^f]^{1/2}. \qquad (8.50)$$

The properties of $\mathbf{G}_N(z)$ will be studied in later sections.

8.8   THE FIR LATTICE STRUCTURE FOR VECTOR LPC

The update equations for predictor polynomials, given in Eqs. (8.37(f)) and (8.37(g)), can be used to derive lattice structures for vector linear prediction, similar to the scalar case. Recall that the error $\mathbf{e}_m^f(n)$ is the output of $\mathbf{A}_m(z)$ in response to the input $\mathbf{x}(n)$, and similarly, $\mathbf{e}_m^{\mathrm{b}}(n)$ is the output of $\mathbf{B}_m(z)$. Thus, the update Eqs. (8.37(f)) and (8.37(g)) show that the prediction errors for order $m + 1$ can be derived from those for order m using the MIMO lattice section shown in Fig. 8.2.

Comparing this with the lattice structure for the scalar case (Fig. 4.5), we see that there is a great deal of similarity. One difference is that, in the scalar case, the lattice coefficients were k_{m+1} and its conjugate k_{m+1}^*. But in the MIMO case, the lattice coefficients are the two different matrices $\mathbf{K}_{m+1}^f$ and $(\mathbf{K}_{m+1}^{\mathrm{b}})^\dagger$.

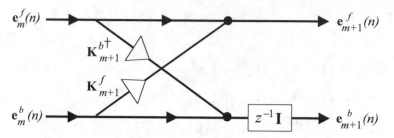

FIGURE 8.2: The MIMO lattice section that generates prediction errors for order $m + 1$ from prediction errors for order m.

8.8.1   Toward Rearrangement of the Lattice

We now show how to bring more symmetry into the lattice structure by rewriting equations a little bit. Recall that the coefficients $\mathbf{K}^f_{m+1}$ and $\mathbf{K}^b_{m+1}$ are related as in Eq. (8.42). We first rearrange this. Because error covariances are positive semidefinite matrices, they can be factored into the form

$$\mathbf{E} = \mathbf{E}^{1/2}\mathbf{E}^{\dagger/2},$$

where $\mathbf{E}^{1/2}$ can be regarded as the left 'square root and its transpose conjugate $\mathbf{E}^{\dagger/2}$ the right square root. Using this notation and remembering that the error covariances are assumed to be nonsingular, we can rewrite Eq. (8.42) as

$$(\mathbf{E}^f_m)^{-1/2}\mathbf{K}^f_{m+1}(\mathbf{E}^b_m)^{1/2} = (\mathbf{E}^f_m)^{\dagger/2}\mathbf{K}^b_{m+1}(\mathbf{E}^b_m)^{-\dagger/2}.$$

We will indicate this quantity by the notation $\mathbf{P}_{m+1}$, that is,

$$\mathbf{P}_{m+1} \triangleq (\mathbf{E}^f_m)^{-1/2}\mathbf{K}^f_{m+1}(\mathbf{E}^b_m)^{1/2} = (\mathbf{E}^f_m)^{\dagger/2}\mathbf{K}^b_{m+1}(\mathbf{E}^b_m)^{-\dagger/2}, \qquad (8.51)$$

so that

$$\mathbf{K}^f_{m+1} = (\mathbf{E}^f_m)^{1/2}\mathbf{P}_{m+1}(\mathbf{E}^b_m)^{-1/2} \quad \text{and} \quad \mathbf{K}^b_{m+1} = (\mathbf{E}^f_m)^{-\dagger/2}\mathbf{P}_{m+1}(\mathbf{E}^b_m)^{\dagger/2}. \qquad (8.52)$$

From the definition of $\mathbf{P}_{m+1}$, we see that

$$\mathbf{P}_{m+1}\mathbf{P}^\dagger_{m+1} = (\mathbf{E}^f_m)^{-1/2}(\mathbf{K}^f_{m+1})(\mathbf{K}^b_{m+1})^\dagger(\mathbf{E}^f_m)^{1/2} \qquad (8.53)$$

and

$$\mathbf{P}^\dagger_{m+1}\mathbf{P}_{m+1} = (\mathbf{E}^b_m)^{-1/2}(\mathbf{K}^b_{m+1})^\dagger\mathbf{K}^f_{m+1}(\mathbf{E}^b_m)^{1/2}. \qquad (8.54)$$

To understand the role of these matrices, recall the error update Eq. (8.35). This can be rewritten as

$$
\begin{aligned}
\mathbf{E}_{m+1}^{f} &= \left(\mathbf{I}_L - \mathbf{K}_{m+1}^{f}(\mathbf{K}_{m+1}^{b})^{\dagger}\right)\mathbf{E}_m^{f} \\
&= \left(\mathbf{I}_L - \mathbf{K}_{m+1}^{f}(\mathbf{K}_{m+1}^{b})^{\dagger}\right)(\mathbf{E}_m^{f})^{1/2}(\mathbf{E}_m^{f})^{\dagger/2} \\
&= \left((\mathbf{E}_m^{f})^{1/2} - \mathbf{K}_{m+1}^{f}(\mathbf{K}_{m+1}^{b})^{\dagger}(\mathbf{E}_m^{f})^{1/2}\right)(\mathbf{E}_m^{f})^{\dagger/2} \\
&= (\mathbf{E}_m^{f})^{1/2}\left(\mathbf{I}_L - (\mathbf{E}_m^{f})^{-1/2}\mathbf{K}_{m+1}^{f}(\mathbf{K}_{m+1}^{b})^{\dagger}(\mathbf{E}_m^{f})^{1/2}\right)(\mathbf{E}_m^{f})^{\dagger/2},
\end{aligned}
$$

so that

$$
\mathbf{E}_{m+1}^{f} = (\mathbf{E}_m^{f})^{1/2}\left(\mathbf{I}_L - \mathbf{P}_{m+1}\mathbf{P}_{m+1}^{\dagger}\right)(\mathbf{E}_m^{f})^{\dagger/2}. \tag{8.55}
$$

Similarly, from Eq. (8.36), we have

$$
\begin{aligned}
\mathbf{E}_{m+1}^{b} &= \left(\mathbf{I}_L - (\mathbf{K}_{m+1}^{b})^{\dagger}\mathbf{K}_{m+1}^{f}\right)\mathbf{E}_m^{b} \\
&= \left(\mathbf{I}_L - (\mathbf{K}_{m+1}^{b})^{\dagger}\mathbf{K}_{m+1}^{f}\right)(\mathbf{E}_m^{b})^{1/2}(\mathbf{E}_m^{b})^{\dagger/2} \\
&= \left((\mathbf{E}_m^{b})^{1/2} - (\mathbf{K}_{m+1}^{b})^{\dagger}\mathbf{K}_{m+1}^{f}(\mathbf{E}_m^{b})^{1/2}\right)(\mathbf{E}_m^{b})^{\dagger/2} \\
&= (\mathbf{E}_m^{b})^{1/2}\left(\mathbf{I}_L - (\mathbf{E}_m^{b})^{-1/2}(\mathbf{K}_{m+1}^{b})^{\dagger}\mathbf{K}_{m+1}^{f}(\mathbf{E}_m^{b})^{1/2}\right)(\mathbf{E}_m^{b})^{\dagger/2},
\end{aligned}
$$

so that

$$
\mathbf{E}_{m+1}^{b} = (\mathbf{E}_m^{b})^{1/2}\left(\mathbf{I}_L - \mathbf{P}_{m+1}^{\dagger}\mathbf{P}_{m+1}\right)(\mathbf{E}_m^{b})^{\dagger/2}. \tag{8.56}
$$

Thus, the modified lattice coefficient $\mathbf{P}_{m+1}$ enters both the forward and backward error updates in a symmetrical way. Under the assumption that the error covariances are nonsingular, we now prove that $\mathbf{P}_{m+1}$ satisfies a boundedness property:

$$
\mathbf{P}_{m+1}\mathbf{P}_{m+1}^{\dagger} < \mathbf{I} \tag{8.57}
$$

and similarly

$$
\mathbf{P}_{m+1}^{\dagger}\mathbf{P}_{m+1} < \mathbf{I}. \tag{8.58}
$$

These can equivalently be written as

$$
(\mathbf{I} - \mathbf{P}_{m+1}\mathbf{P}_{m+1}^{\dagger}) > 0, \quad (\mathbf{I} - \mathbf{P}_{m+1}^{\dagger}\mathbf{P}_{m+1}) > 0. \tag{8.59}
$$

Proof. Eq. (8.55) can be written as

$$
\mathbf{E}_{m+1}^{f} = \mathbf{U}^{\dagger}\mathbf{Q}\mathbf{U},
$$

where $\mathbf{Q}$ is Hermitian and $\mathbf{U}$ is nonsingular. Because $\mathbf{E}^f_{m+1}$ is assumed to be nonsingular, it is positive definite, so that $\mathbf{v}^\dagger \mathbf{E}^f_{m+1}\mathbf{v} > 0$ for any nonzero vector $\mathbf{v}$. Thus, $\mathbf{v}^\dagger \mathbf{U}^\dagger \mathbf{Q}\mathbf{U}\mathbf{v} > 0$. Because $\mathbf{U} = (\mathbf{E}^f_m)^{\dagger/2}$ is nonsingular, this implies that $\mathbf{w}^\dagger \mathbf{Q}\mathbf{w} > 0$ for any vector $\mathbf{w} \neq \mathbf{0}$, which is equivalent to the statement that $\mathbf{Q}$ is positive definite. This proves that $\mathbf{I} - \mathbf{P}_{m+1}\mathbf{P}^\dagger_{m+1}$ is positive definite, which is equivalent to Eq. (8.57). Eq. (8.58) follows similarly. $\qquad\square$

Because $\mathbf{E}^f_{m+1}$ is Hermitian and positive definite, it can be written in the form

$$\mathbf{E}^f_{m+1} = (\mathbf{E}^f_{m+1})^{1/2}(\mathbf{E}^f_{m+1})^{\dagger/2}$$

From Eq. (8.55), we therefore see that the left square root can be expressed as

$$(\mathbf{E}^f_{m+1})^{1/2} = (\mathbf{E}^f_m)^{1/2}\left(\mathbf{I}_L - \mathbf{P}_{m+1}\mathbf{P}^\dagger_{m+1}\right)^{1/2}. \qquad (8.60)$$

The existence of the square root $(\mathbf{I}_L - \mathbf{P}_{m+1}\mathbf{P}^\dagger_{m+1})^{1/2}$ is guaranteed by the fact that $(\mathbf{I}_L - \mathbf{P}_{m+1}\mathbf{P}^\dagger_{m+1})$ is positive definite (as shown by Eq. (8.59)). Similarly, from Eq. (8.56), we have

$$(\mathbf{E}^b_{m+1})^{1/2} = (\mathbf{E}^b_m)^{1/2}\left(\mathbf{I}_L - \mathbf{P}^\dagger_{m+1}\mathbf{P}_{m+1}\right)^{1/2} \qquad (8.61)$$

8.8.2 The Symmetrical Lattice

The advantage of defining the $L \times L$ matrix $\mathbf{P}_{m+1}$ from the $L \times L$ matrices $\mathbf{K}^f_{m+1}$ and $\mathbf{K}^b_{m+1}$ is that it allows us to redraw the lattice more symmetrically, in terms of $\mathbf{P}_{m+1}$. Because $\mathbf{P}_{m+1}$ also has the boundedness property (8.57), the lattice becomes very similar to the scalar LPC lattice, which had the boundedness property $|k_m| < 1$. We now derive this symmetrical lattice. From Eq. (8.51), we see that the lattice coefficients can be expressed as

$$\mathbf{K}^f_{m+1} = (\mathbf{E}^f_m)^{1/2}\mathbf{P}_{m+1}(\mathbf{E}^b_m)^{-1/2}, \qquad (8.62)$$

and

$$\mathbf{K}^b_{m+1} = (\mathbf{E}^f_m)^{-\dagger/2}\mathbf{P}_{m+1}(\mathbf{E}^b_m)^{\dagger/2}. \qquad (8.63)$$

Substituting these into Fig. 8.2 and rearranging, we obtain the lattice section shown in Fig. 8.3(b). The quantities $\mathbf{g}^f_m(n)$ and $\mathbf{g}^b_m(n)$ indicated in Fig. 8.3(b) are given by

$$\mathbf{g}^f_m(n) = (\mathbf{E}^f_m)^{-1/2}\mathbf{e}^f_m(n), \quad \mathbf{g}^b_m(n) = (\mathbf{E}^b_m)^{-1/2}\mathbf{e}^b_m(n) \qquad (8.64)$$

These are *normalized prediction error* vectors, in the sense that their covariance matrix is identity. We can generate the set of all forward and backward normalized error sequences by using the cascaded FIR lattice structure, as demonstrated in Fig. 8.4 for three stages.

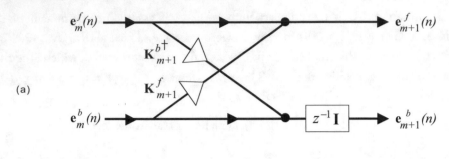

(a)

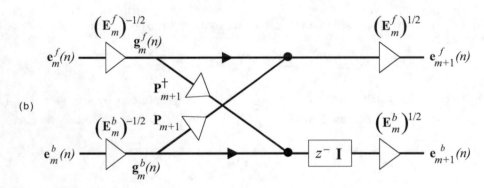

(b)

FIGURE 8.3: (a) The original MIMO lattice section and (b) an equivalent section in a more symmetrical form.

The multipliers $\mathbf{Q}_m^f$ in the figure are given by

$$\mathbf{Q}_m^f = (\mathbf{E}_m^f)^{-1/2}(\mathbf{E}_{m-1}^f)^{1/2}, \quad m \geq 1 \qquad (8.65)$$

and $\mathbf{Q}_0^f = (\mathbf{E}_0^f)^{-1/2}$. The multipliers $\mathbf{Q}_m^b$ are similar (just replace superscript f with b everywhere). In view of Eqs. (8.60) and (8.61), these multipliers can be rewritten as

$$\mathbf{Q}_m^f = \left(\mathbf{I}_L - \mathbf{P}_m\mathbf{P}_m^\dagger\right)^{-1/2}, \quad \mathbf{Q}_m^b = \left(\mathbf{I}_L - \mathbf{P}_m^\dagger\mathbf{P}_m\right)^{-1/2} \quad m \geq 1. \qquad (8.66)$$

The unnormalized error sequence for any order can be obtained simply by inserting matrix multipliers $(\mathbf{E}_m^f)^{1/2}$ and $(\mathbf{E}_m^b)^{1/2}$ at appropriate places. Because

$$\mathbf{e}_3^f(n) = (\mathbf{E}_3^f)^{1/2}\mathbf{g}_3^f(n) \quad \text{and} \quad \mathbf{e}_3^b(n) = (\mathbf{E}_3^b)^{1/2}\mathbf{g}_3^b(n),$$

the extra (rightmost) multipliers in Fig. 8.4(a) are therefore $(\mathbf{E}_3^f)^{1/2}$ and $(\mathbf{E}_3^b)^{1/2}$. The lattice sections in Fig. 8.4 will be referred to as *normalized lattice* sections because the outputs at each of the stages represent the normalized error sequences.

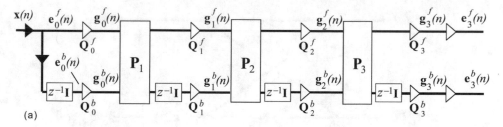

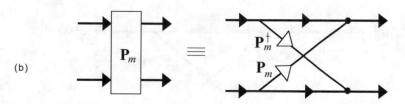

FIGURE 8.4: (a) The MIMO FIR LPC lattice structure with normalized sections and (b) details of the box labeled $\mathbf{P}_m$.

8.9   THE IIR LATTICE STRUCTURE FOR VECTOR LPC

The IIR LPC lattice for the vector case can be obtained similar to the scalar case by starting from Fig. 8.2 reproduced in Fig. 8.5(a) and rearranging equations. From Fig. 8.5(a), we have

$$\mathbf{e}_{m+1}^{f}(n) = \mathbf{e}_{m}^{f}(n) + \mathbf{K}_{m+1}^{f}\mathbf{e}_{m}^{b}(n)$$

and

$$\mathbf{e}_{m+1}^{b}(n+1) = [\mathbf{K}_{m+1}^{b}]^{\dagger}\mathbf{e}_{m}^{f}(n) + \mathbf{e}_{m}^{b}(n).$$

The first equation can be rewritten as

$$\mathbf{e}_{m}^{f}(n) = \mathbf{e}_{m+1}^{f}(n) - \mathbf{K}_{m+1}^{f}\mathbf{e}_{m}^{b}(n).$$

Substituting this into the second equation, we get

$$\mathbf{e}_{m+1}^{b}(n+1) = [\mathbf{K}_{m+1}^{b}]^{\dagger}\mathbf{e}_{m+1}^{f}(n) + \left(\mathbf{I} - [\mathbf{K}_{m+1}^{b}]^{\dagger}\mathbf{K}_{m+1}^{f}\right)\mathbf{e}_{m}^{b}(n).$$

These equations can be "drawn" as shown in Fig. 8.5(b). By repeated application of this idea, we obtain the IIR LPC lattice shown in Fig. 8.6. This is called the MIMO IIR LPC lattice. The transfer functions indicated as $\mathbf{H}_k(z)$ in the figure will be discussed later.

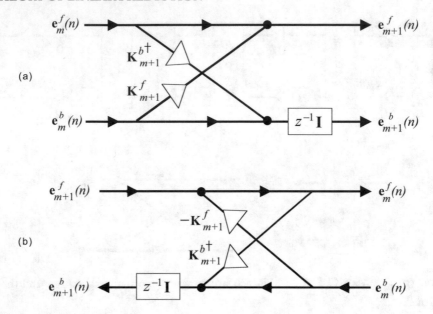

FIGURE 8.5: (a) The MIMO FIR lattice section and (b) the corresponding IIR lattice section.

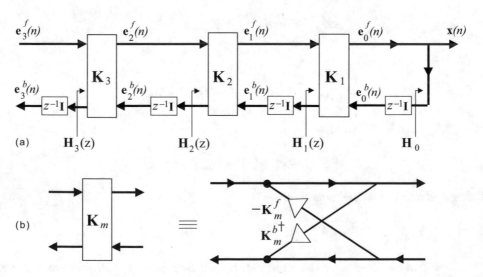

FIGURE 8.6: (a) The MIMO IIR cascaded LPC lattice, and (b) details of the building block labeled $\mathbf{K}_m$.

8.10   THE NORMALIZED IIR LATTICE

Starting from the symmetrical MIMO FIR LPC lattice in Fig. 8.3(b), we can derive an IIR LPC lattice with slightly different structural arrangement. We will call this the normalized IIR LPC lattice for reasons that will become clear. To obtain this structure, consider again the MIMO FIR lattice section shown separately in Fig. 8.7(a). This FIR lattice section can be represented using the equations

$$\mathbf{g}_{m+1}^{f}(n) = \mathbf{Q}_{m+1}^{f}\left(\mathbf{g}_{m}^{f}(n) + \mathbf{P}_{m+1}\mathbf{g}_{m}^{b}(n)\right) \tag{8.67}$$

and

$$\mathbf{g}_{m+1}^{b}(n+1) = \mathbf{Q}_{m+1}^{b}\left(\mathbf{P}_{m+1}^{\dagger}\mathbf{g}_{m}^{f}(n) + \mathbf{g}_{m}^{b}(n)\right). \tag{8.68}$$

The first equation can be rearranged to express $\mathbf{g}_{m}^{f}(n)$ in terms of the other quantities:

$$\mathbf{g}_{m}^{f}(n) = (\mathbf{Q}_{m+1}^{f})^{-1}\mathbf{g}_{m+1}^{f}(n) - \mathbf{P}_{m+1}\mathbf{g}_{m}^{b}(n). \tag{8.69}$$

The second equation (8.68) can then be rewritten by substituting from Eq. (8.69):

$$\mathbf{g}_{m+1}^{b}(n+1) = \mathbf{Q}_{m+1}^{b}\mathbf{P}_{m+1}^{\dagger}(\mathbf{Q}_{m+1}^{f})^{-1}\mathbf{g}_{m+1}^{f}(n) + (\mathbf{Q}_{m+1}^{b})^{-\dagger}\mathbf{g}_{m}^{b}(n), \tag{8.70}$$

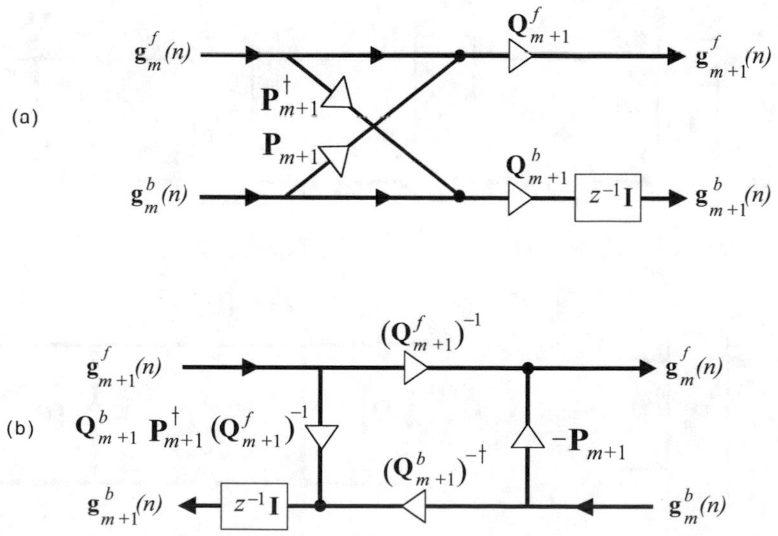

FIGURE 8.7: (a) The normalized MIMO FIR lattice section and (b) the corresponding IIR normalized lattice section.

where we have used the definitions (8.66) for $\mathbf{Q}_m^f$ and $\mathbf{Q}_m^b$, that is,

$$\mathbf{Q}_m^f = \left(\mathbf{I}_L - \mathbf{P}_m\mathbf{P}_m^\dagger\right)^{-1/2}, \quad \mathbf{Q}_m^b = \left(\mathbf{I}_L - \mathbf{P}_m^\dagger\mathbf{P}_m\right)^{-1/2} \quad m \geq 1.$$

The preceding two equations can be represented using the normalized IIR lattice section shown in Fig. 8.7(b).

The normalized MIMO IIR LPC lattice therefore takes the form shown in Fig. 8.8 for three sections. For an N-stage IIR lattice, the signal entering at the left is the forward error $\mathbf{e}_N^f(n)$, and the signal at the right end is $\mathbf{g}_0^f(n)$. Upon scaling with the multiplier

$$(\mathbf{Q}_0^f)^{-1} = (\mathbf{E}_0^f)^{1/2},$$

the error signal $\mathbf{g}_0^f(n)$ becomes $\mathbf{e}_0^f(n)$, which is nothing but the original signal $\mathbf{x}(n)$. Thus, while the FIR lattice produced the (normalized) error signals from the original signal $\mathbf{x}(n)$, the IIR lattice reproduces $\mathbf{x}(n)$ back from the error signals.

8.11　THE PARAUNITARY OR MIMO ALL-PASS PROPERTY

Figure 8.9 shows one section of the IIR lattice structure. Here, $\mathbf{G}_m(z)$ is the transfer function from the pervious section, and $\mathbf{P}_{m+1}$ is the constant matrix shown in Fig. 8.8(b). Comparing with Fig. 8.8(a), we see that the transfer matrices $\mathbf{G}_m(z)$ convert the forward error signals to corresponding

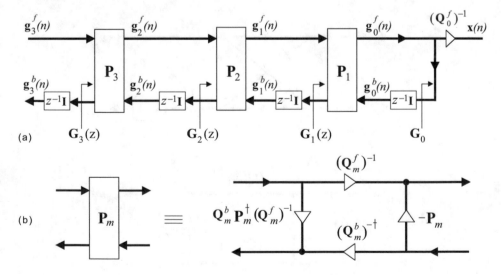

FIGURE 8.8: (a) The normalized MIMO IIR LPC lattice shown for three stages and (b) details of the normalized IIR lattice sections. Here, $\mathbf{Q}_m^f = (\mathbf{I}_L - \mathbf{P}_m\mathbf{P}_m^\dagger)^{-1/2}$, $\mathbf{Q}_m^b = (\mathbf{I}_L - \mathbf{P}_m^\dagger\mathbf{P}_m)^{-1/2}$, $m \geq 1$, and $\mathbf{Q}_0^f = (\mathbf{E}_0^f)^{-1/2}$.

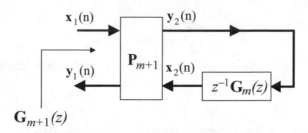

FIGURE 8.9: The normalized MIMO IIR lattice section.

backward error signals. More precisely, if $\mathbf{g}_m^f(n)$ is input to the system $\mathbf{G}_m(z)$, then the output is $\mathbf{g}_m^b(n+1)$.

In the scalar case, the transfer function $G_m(z)$ was all-pass, and this was used to show that $G_{m+1}(z)$ was all-pass. Furthermore, these all-pass filters had all their poles inside the unit circle (Section 4.3.2). We now prove an analogous property for the vector case. This is somewhat more complicated than in the scalar case because the transfer functions are MIMO, and the all-pass property generalizes to the so-called paraunitary property to be defined below (Vaidyanathan, 1993). We will first show that the building block indicated as $\mathbf{P}_{m+1}$ has a unitary transfer matrix $\mathbf{T}_{m+1}$. We then show that in an interconnection such as this, whenever $\mathbf{G}_m(z)$ is paraunitary and $\mathbf{P}_{m+1}$ unitary, then $\mathbf{G}_{m+1}(z)$ is paraunitary as well. Because $\mathbf{G}_0(z) = \mathbf{I}$ is paraunitary, it will therefore follow that all the transfer functions $\mathbf{G}_m(z)$ in the normalized IIR lattice are paraunitary. At this point, we ask the reader to review the tilde notation $\widetilde{\mathbf{G}}(z)$ described in Section 1.2.1.

Definition 8.1. *Paraunitary Systems.* An $M \times M$ transfer matrix $\mathbf{G}(z)$ is said to be paraunitary if

$$\mathbf{G}^\dagger(e^{j\omega})\mathbf{G}(e^{j\omega}) = \mathbf{I} \qquad (8.71)$$

for all ω. That is, the transfer matrix is unitary for all frequencies. If $\mathbf{G}(z)$ is rational (i.e., the entries $G_{km}(z)$ are ratios of polynomials in z^{-1}) then the paraunitary property is equivalent to

$$\widetilde{\mathbf{G}}(z)\mathbf{G}(z) = \mathbf{I} \qquad (8.72)$$

for all z. $\Diamond$

With $\mathbf{x}(n)$ and $\mathbf{y}(n)$ denoting the input and output of an LTI system (Fig. 8.10), their Fourier transforms are related as

$$\mathbf{Y}(e^{j\omega}) = \mathbf{G}(e^{j\omega})\mathbf{X}(e^{j\omega})$$

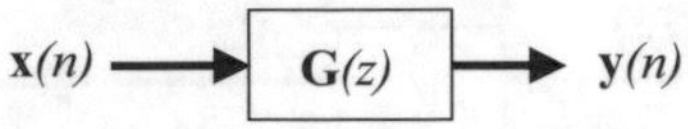

$$x(n) \longrightarrow \boxed{G(z)} \longrightarrow y(n)$$

FIGURE 8.10: An LTI system with input $x(n)$ and output $y(n)$.

A number of important properties of paraunitary matrices are derived in Vaidyanathan (1993). The following properties are especially relevant for our discussions here:

1. If $\mathbf{G}(e^{j\omega})$ is unitary, it follows that

$$\mathbf{Y}^\dagger(e^{j\omega})\mathbf{Y}(e^{j\omega}) = \mathbf{X}^\dagger(e^{j\omega})\mathbf{X}(e^{j\omega}), \quad \text{for all } \omega \qquad (8.73)$$

   for any Fourier transformable input $x(n)$.

2. It can, in fact, be shown that a rational LTI system $\mathbf{G}(z)$ is paraunitary if and only if (8.73) holds for all Fourier-transformable inputs $x(n)$.

3. By integrating both sides of (8.73) and using Parseval's relation, it also follows that, for a paraunitary system, the input energy is equal to the output energy, that is,

$$\sum_n \mathbf{x}^\dagger(n)\mathbf{x}(n) = \sum_n \mathbf{y}^\dagger(n)\mathbf{y}(n) \qquad (8.74)$$

   for all finite energy inputs.

8.11.1   Unitarity of Building Blocks

The building block $\mathbf{P}_{m+1}$ in Fig. 8.9 has the form shown in Fig. 8.8(b). Its transfer matrix has the form

$$\mathbf{T} = \begin{bmatrix} \mathbf{Q}^b \mathbf{P}^\dagger (\mathbf{Q}^f)^{-1} & (\mathbf{Q}^b)^{-\dagger} \\ (\mathbf{Q}^f)^{-1} & -\mathbf{P} \end{bmatrix}, \qquad (8.75)$$

where all subscripts have been omitted for simplicity. Recall here that the matrices $\mathbf{Q}^b$ and $\mathbf{Q}^f$ are related to $\mathbf{P}$ as in Eq. (8.66), that is,

$$\mathbf{Q}^f = \left(\mathbf{I}_L - \mathbf{P}\mathbf{P}^\dagger\right)^{-1/2}, \quad \mathbf{Q}^b = \left(\mathbf{I}_L - \mathbf{P}^\dagger\mathbf{P}\right)^{-1/2}. \qquad (8.76)$$

We will now show that $\mathbf{T}$ is unitary, that is, $\mathbf{T}^\dagger\mathbf{T} = \mathbf{I}$.

Proof of Unitarity of T. We have

$$\mathbf{T}^\dagger \mathbf{T} = \begin{bmatrix} \mathbf{A} & \mathbf{B} \\ \mathbf{B}^\dagger & \mathbf{D} \end{bmatrix},$$

where the symbols above are defined as

$$\mathbf{A} = (\mathbf{Q}^f)^{-\dagger}\mathbf{P}(\mathbf{Q}^b)^\dagger \mathbf{Q}^b \mathbf{P}^\dagger (\mathbf{Q}^f)^{-1} + (\mathbf{Q}^f)^{-\dagger}(\mathbf{Q}^f)^{-1},$$
$$\mathbf{B} = (\mathbf{Q}^f)^{-\dagger}\mathbf{P}(\mathbf{Q}^b)^\dagger (\mathbf{Q}^b)^{-\dagger} - (\mathbf{Q}^f)^{-\dagger}\mathbf{P} = \mathbf{0},$$
$$\mathbf{D} = (\mathbf{Q}^b)^{-1}(\mathbf{Q}^b)^{-\dagger} + \mathbf{P}^\dagger \mathbf{P}.$$

So, **B** is trivially zero. To simplify **D**, we substitute from Eq. (8.76) to obtain

$$\mathbf{D} = \left(\mathbf{I}_L - \mathbf{P}^\dagger \mathbf{P}\right)^{1/2} \left(\mathbf{I}_L - \mathbf{P}^\dagger \mathbf{P}\right)^{\dagger/2} + \mathbf{P}^\dagger \mathbf{P}$$
$$= \left(\mathbf{I}_L - \mathbf{P}^\dagger \mathbf{P}\right) + \mathbf{P}^\dagger \mathbf{P} = \mathbf{I}_L$$

Next, substituting from Eq. (8.76), the first term of **A** becomes

$$\left(\mathbf{I}_L - \mathbf{P}\mathbf{P}^\dagger\right)^{\dagger/2} \mathbf{P} \left(\mathbf{I}_L - \mathbf{P}^\dagger\mathbf{P}\right)^{-\dagger/2} \left(\mathbf{I}_L - \mathbf{P}^\dagger\mathbf{P}\right)^{-1/2} \mathbf{P}^\dagger \left(\mathbf{I}_L - \mathbf{P}\mathbf{P}^\dagger\right)^{1/2}$$
$$= \left(\mathbf{I}_L - \mathbf{P}\mathbf{P}^\dagger\right)^{\dagger/2} \mathbf{P} \left(\mathbf{I}_L - \mathbf{P}^\dagger\mathbf{P}\right)^{-1} \mathbf{P}^\dagger \left(\mathbf{I}_L - \mathbf{P}\mathbf{P}^\dagger\right)^{1/2},$$

so that

$$\mathbf{A} = \left(\mathbf{I}_L - \mathbf{P}\mathbf{P}^\dagger\right)^{\dagger/2} \underbrace{\left(\mathbf{P} \left(\mathbf{I}_L - \mathbf{P}^\dagger\mathbf{P}\right)^{-1} \mathbf{P}^\dagger + \mathbf{I}\right)}_{\text{call this } \mathbf{A}_1} \left(\mathbf{I}_L - \mathbf{P}\mathbf{P}^\dagger\right)^{1/2}.$$

We now claim that $\mathbf{A}_1 = (\mathbf{I}_L - \mathbf{P}\mathbf{P}^\dagger)^{-1}$:

$$\mathbf{A}_1(\mathbf{I}_L - \mathbf{P}\mathbf{P}^\dagger) = \mathbf{P}\left(\mathbf{I}_L - \mathbf{P}^\dagger\mathbf{P}\right)^{-1}\left(\mathbf{P}^\dagger - \mathbf{P}^\dagger\mathbf{P}\mathbf{P}^\dagger\right) + \left(\mathbf{I}_L - \mathbf{P}\mathbf{P}^\dagger\right)$$
$$= \mathbf{P}\left(\mathbf{I}_L - \mathbf{P}^\dagger\mathbf{P}\right)^{-1}\left(\mathbf{I}_L - \mathbf{P}^\dagger\mathbf{P}\right)\mathbf{P}^\dagger + \left(\mathbf{I}_L - \mathbf{P}\mathbf{P}^\dagger\right)$$
$$= \mathbf{P}\mathbf{P}^\dagger + \left(\mathbf{I}_L - \mathbf{P}\mathbf{P}^\dagger\right) = \mathbf{I}_L.$$

Thus, $\mathbf{A} = (\mathbf{I}_L - \mathbf{P}\mathbf{P}^\dagger)^{\dagger/2}(\mathbf{I}_L - \mathbf{P}\mathbf{P}^\dagger)^{-1}(\mathbf{I}_L - \mathbf{P}\mathbf{P}^\dagger)^{1/2} = \mathbf{I}$, where we have used $(\mathbf{I}_L - \mathbf{P}\mathbf{P}^\dagger) = (\mathbf{I}_L - \mathbf{P}\mathbf{P}^\dagger)^{1/2}(\mathbf{I}_L - \mathbf{P}\mathbf{P}^\dagger)^{\dagger/2}$. This completes the proof that **T** is unitary. □

8.11.2 Propagation of Paraunitary Property

Return now to Fig. 8.9. We will show that if the rational transfer matrix $\mathbf{G}_m(z)$ is paraunitary and the box labeled $\mathbf{P}_{m+1}$ has a unitary transfer matrix $\mathbf{T}_{m+1}(z)$ then $\mathbf{G}_{m+1}(z)$ is also paraunitary.

Proof of Paraunitarity of $\mathbf{G}_{m+1}(z)$. From Fig. 8.9, we have

$$\begin{bmatrix} \mathbf{Y}_1(e^{j\omega}) \\ \mathbf{Y}_2(e^{j\omega}) \end{bmatrix} = \mathbf{T}_{m+1} \begin{bmatrix} \mathbf{X}_1(e^{j\omega}) \\ \mathbf{X}_2(e^{j\omega}) \end{bmatrix}$$

The unitary property of $\mathbf{T}_{m+1}$ implies that

$$\mathbf{Y}_1^\dagger(e^{j\omega})\mathbf{Y}_1(e^{j\omega}) + \mathbf{Y}_2^\dagger(e^{j\omega})\mathbf{Y}_2(e^{j\omega})$$

$$= \mathbf{X}_1^\dagger(e^{j\omega})\mathbf{X}_1(e^{j\omega}) + \mathbf{X}_2^\dagger(e^{j\omega})\mathbf{X}_2(e^{j\omega}), \quad \text{for all } \omega.$$

Because $\mathbf{G}_m(z)$ is paraunitary, we also have $\mathbf{Y}_2^\dagger(e^{j\omega})\mathbf{Y}_2(e^{j\omega}) = \mathbf{X}_2^\dagger(e^{j\omega})\mathbf{X}_2(e^{j\omega})$ for all ω. So it follows that

$$\mathbf{Y}_1^\dagger(e^{j\omega})\mathbf{Y}_1(e^{j\omega}) = \mathbf{X}_1^\dagger(e^{j\omega})\mathbf{X}_1(e^{j\omega}), \quad \text{for all } \omega.$$

This shows that the rational function $\mathbf{G}_{m+1}(z)$ is paraunitary. $\square$

8.11.3   Poles of the MIMO IIR Lattice

Figure 8.11 is a reproduction of Fig. 8.9 with details of the building block $\mathbf{P}_{m+1}$ shown explicitly (with subscripts deleted for simplicity). Here, the matrices $\mathbf{Q}^f$ and $\mathbf{Q}^b$ are as in Eq. (8.66), that is,

$$\mathbf{Q}^f = \left(\mathbf{I}_L - \mathbf{P}\mathbf{P}^\dagger\right)^{-1/2} \quad \text{and} \quad \mathbf{Q}^b = \left(\mathbf{I}_L - \mathbf{P}^\dagger\mathbf{P}\right)^{-1/2}.$$

In Section 8.8.1, we showed that the LPC lattice satisfies the property

$$\mathbf{P}^\dagger\mathbf{P} < \mathbf{I} \qquad\qquad (8.77)$$

(see Eq. (8.58)). Under this assumption, we will prove an important property about the poles[3] of the transfer matrix $\mathbf{G}_{m+1}(z)$.

Lemma 8.1. *Poles of the IIR Lattice.* If the rational paraunitary matrix $\mathbf{G}_m(z)$ has all poles in $|z| < 1$ and if the matrix multiplier $\mathbf{P}$ is bounded as in Eq. (8.77), then the poles of $\mathbf{G}_{m+1}(z)$ are also confined to the region $|z| < 1$. $\Diamond$

Proof. It is sufficient to show that the transfer function $\mathbf{F}(z)$ indicated in Fig. 8.11 has all poles in $|z| < 1$. To find an expression for $\mathbf{F}(z)$, note that

$$\mathbf{Y}(z) = z^{-1}\mathbf{G}_m(z)\mathbf{W}(z) \quad \text{and} \quad \mathbf{W}(z) = -z^{-1}\mathbf{P}\mathbf{G}_m(z)\mathbf{W}(z) + \mathbf{X}(z),$$

---

[3]The poles of a MIMO transfer function $\mathbf{F}(z)$ are nothing but the poles of the individual elements $F_{km}(z)$. For this section, the reader may want to review the theory on poles of MIMO systems (Kailath, 1980; Vaidyanathan, 1993, Chapter 13).

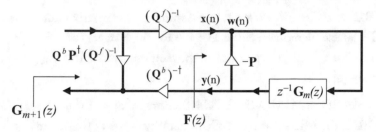

FIGURE 8.11: The normalized MIMO IIR lattice section with some of the details indicated. Subscripts on **P** and **Q** are omitted for simplicity.

where $\mathbf{w}(n)$ is the signal indicated in the figure. Eliminating $\mathbf{W}(z)$, we get $\mathbf{Y}(z) = z^{-1}\mathbf{G}_m(z)(\mathbf{I} + z^{-1}\mathbf{P}\mathbf{G}_m(z))^{-1}\mathbf{X}(z)$. This shows that

$$\mathbf{F}(z) = z^{-1}\mathbf{G}_m(z)(\mathbf{I} + z^{-1}\mathbf{P}\mathbf{G}_m(z))^{-1}. \tag{8.78}$$

Because the poles of $\mathbf{G}_m(z)$ are assumed to be in $|z| < 1$, it suffices to show that all the poles of $(\mathbf{I} + z^{-1}\mathbf{P}\mathbf{G}_m(z))^{-1}$ are in $|z| < 1$. Because

$$\left(\mathbf{I} + z^{-1}\mathbf{P}\mathbf{G}_m(z)\right)^{-1} = \frac{\text{Adj}\left(\mathbf{I} + z^{-1}\mathbf{P}\mathbf{G}_m(z)\right)}{\det\left(\mathbf{I} + z^{-1}\mathbf{P}\mathbf{G}_m(z)\right)},$$

where Adj is the adjugate matrix, it is sufficient to show that all zeros of the denominator above are in $|z| < 1$. We will use Eq. (8.77) and the assumption that $\mathbf{G}_m(z)$ is paraunitary with all poles in $|z| < 1$. The latter implies

$$\mathbf{G}_m^\dagger(z)\mathbf{G}_m(z) \leq \mathbf{I} \tag{8.79}$$

for all z in $|z| \geq 1$ (see Section 14.8 of Vaidyanathan, 1993). Now, if z_0 is a zero of the determinant of $(\mathbf{I} + z^{-1}\mathbf{P}\mathbf{G}_m(z))$, then $\mathbf{I} + z_0^{-1}\mathbf{P}\mathbf{G}_m(z_0)$ is singular. So there exists a unit-norm vector $\mathbf{v}$ such that $z_0^{-1}\mathbf{P}\mathbf{G}_m(z_0)\mathbf{v} = -\mathbf{v}$. This implies

$$\mathbf{v}^\dagger\mathbf{G}_m^\dagger(z_0)\mathbf{P}^\dagger\mathbf{P}\mathbf{G}_m(z_0)\mathbf{v} = |z_0|^2. \tag{8.80}$$

If $|z_0| \geq 1$, then $\mathbf{G}_m(z_0)\mathbf{v}$ has norm ≤ 1 (in view of Eq. (8.79)). So the left-hand side in Eq. (8.80) is < 1 in view of (8.77). This contradicts the statement that the right-hand side of Eq. (8.80) satisfies $|z_0| \geq 1$. This proves that all zeros of the said determinant are in $|z| < 1$. $\square$

By induction, it therefore follows that the MIMO IIR LPC lattice is stable. Summarizing, we have proved the following:

Theorem 8.2. *Stability of the IIR Lattice.* Assume all the matrix multipliers indicated as $\mathbf{P}_m$ in the MIMO IIR lattice (demonstrated in Fig. 8.8 for three stages) satisfy $\mathbf{P}_m^\dagger \mathbf{P}_m < \mathbf{I}$ and define $\mathbf{Q}_m^f$ and $\mathbf{Q}_m^b$ as in (8.66). Then, the poles of all the paraunitary matrices $\mathbf{G}_m(z)$ in the LPC lattice are confined to be in $|z| < 1$. $\diamond$

Equivalently, all poles of the MIMO IIR LPC lattice are in $|z| < 1$ (i.e., the IIR lattice is a causal stable system). This is the MIMO version of the stability property we saw in Section 4.3.2 for the scalar LPC lattice.

8.12 WHITENING EFFECT AND STALLING

Consider again Levinson's recursion for the MIMO linear predictor described in Section 8.5. Recall that the error covariances are updated according to Eqs. (8.37(h)) and (8.37(i)), which can be rewritten using (8.37(d)) and (8.37(e)) as follows:

$$\mathbf{E}_{N+1}^f = \mathbf{E}_N^f - (\mathbf{F}_N^f)^\dagger (\mathbf{E}_N^b)^{-1} \mathbf{F}_N^f \qquad (8.81)$$

$$\mathbf{E}_{N+1}^b = \mathbf{E}_N^b - \mathbf{F}_N^f (\mathbf{E}_N^f)^{-1} (\mathbf{F}_N^f)^\dagger. \qquad (8.82)$$

We say that there is no progress as we move from order N to order $N+1$, if $\mathcal{E}_{N+1}^f = \mathcal{E}_N^f$, that is,

$$\mathrm{Tr}\,\mathbf{E}_{N+1}^f = \mathrm{Tr}\,\mathbf{E}_N^f. \qquad (8.83)$$

We know that the optimal prediction $\mathbf{e}_N^f(n)$ at time n is orthogonal to the samples $\mathbf{x}(n-k)$, for $1 \le k \le N$. We now claim the following:

Lemma 8.2. *No-Progress Situation.* The condition (8.83) arises if and only if

$$(\mathbf{F}_N^f)^\dagger \stackrel{\Delta}{=} E[\mathbf{e}_N^f(n)\mathbf{x}^\dagger(n-N-1)] = \mathbf{0} \qquad (8.84)$$

that is, the error $\mathbf{e}_N^f(n)$ is orthogonal to $\mathbf{x}(n-N-1)$ in addition to being orthogonal to $\mathbf{x}(n-1), \ldots, \mathbf{x}(n-N)$. $\diamond$

Proof. Eq. (8.83) is equivalent to $\mathrm{Tr}\,((\mathbf{F}_N^f)^\dagger (\mathbf{E}_N^b)^{-1} \mathbf{F}_N^f) = 0$, as seen from Eq. (8.45). Because the matrix $(\mathbf{F}_N^f)^\dagger (\mathbf{E}_N^b)^{-1} \mathbf{F}_N^f$ is Hermitian and positive semidefinite, the zero-trace condition is equivalent to the statement[4]

$$(\mathbf{F}_N^f)^\dagger (\mathbf{E}_N^b)^{-1} (\mathbf{F}_N^f) = \mathbf{0}.$$

---

[4]The trace of a matrix $\mathbf{A}$ is the sum of its eigenvalues. When $\mathbf{A}$ is positive semidefinite, all eigenvalues are nonnegative. So, zero-trace implies all eigenvalues are zero. This implies that the matrix itself is zero (because any positive semidefinite matrix can be written as $\mathbf{A} = \mathbf{U}\mathbf{\Lambda}\mathbf{U}^\dagger$ where $\mathbf{\Lambda}$ is the diagonal matrix of eigenvalues and $\mathbf{U}$ unitary).

This is equivalent to $\mathbf{F}_N^f = 0$ (because $(\mathbf{E}_N^b)^{-1}$ is Hermitian and nonsingular). This completes the proof.   □

A number of points should be noted.

1. From Eqs. (8.82) and (8.84), we see that when there is no progress in forward prediction, then there is no progress in backward prediction either (and vice versa).
2. When Eq. (8.84) holds, we have $\mathbf{K}_{N+1}^f = 0$ (from (8.26)) so that $\mathbf{A}_{N+1}(z) = \mathbf{A}_N(z)$. Similarly $\mathbf{K}_{N+1}^b = \mathbf{0}$ and $\mathbf{B}_{N+1}(z) = z^{-1}\mathbf{B}_N(z)$.

We say that linear prediction "stalls" at order N if there is no further progress after stage N, that is, if

$$\mathcal{E}_N^f = \mathcal{E}_{N+1}^f = \mathcal{E}_{N+2}^f = \dots. \qquad (8.85)$$

We now prove the following:

Lemma 8.3. *Stalling and Whitening.* The stalling situation (8.85) occurs if and only if

$$E[\mathbf{e}_N^f(n)(\mathbf{e}_N^f)^\dagger(n-m)] = \mathbf{E}_N^f \delta(m), \qquad (8.86)$$

that is, if and only if $\mathbf{e}_N^f(n)$ is white.   ◊

Proof. From Lemma 8.2, it is clear that stalling arises if and only if

$$\mathbf{F}_m^f = 0, m \geq N,$$

or equivalently,

$$E[\mathbf{e}_N^f(n)\mathbf{x}^\dagger(n-m)] = \mathbf{0}, \quad m \geq N+1.$$

Because this equality holds for $1 \leq m \leq N$ directly because of orthogonality principle, we can say that stalling happens if and only if

$$E[\mathbf{e}_N^f(n)\mathbf{x}^\dagger(n-m)] = \mathbf{0}, \quad m \geq 1. \qquad (8.87)$$

But $\mathbf{e}_N^f(n)$ is a linear combination of $\mathbf{x}(n-m)$, $m \geq 0$, so the preceding equation implies

$$E[\mathbf{e}_N^f(n)(\mathbf{e}_N^f)^\dagger(n-k)] = \mathbf{0}, \quad k \geq 1. \qquad (8.88)$$

Conversely, because $\mathbf{x}(n)$ is a linear combination of $\mathbf{e}_N^f(n-k)$, $k \geq 0$, we see that Eq. (8.88) implies Eq. (8.87) as well. But because (8.88) is equivalent to Eq. (8.86), we have shown that the stalling condition is equivalent to Eq. (8.86).   □

8.13   PROPERTIES OF TRANSFER MATRICES IN LPC THEORY

In the scalar case, the properties of the transfer functions $A_N(z)$ and $B_N(z)$ are well understood. For example, $A_N(z)$ has all zeros strictly inside the unit circle (i.e., it is a minimum-phase polynomial). Furthermore, we showed that

$$B_N(z) = z^{-(N+1)} \widetilde{A}_N(z).$$

This implies that $B_N(z)$ has all its zeros outside the unit circle and the filter $B_N(z)/A_N(z)$ is stable and all-pass. That is, the forward and backward error sequences are related by an all-pass filter, so that $\mathcal{E}_N^f = \mathcal{E}_N^b$. For the case of vector linear prediction, it is trickier to prove similar properties because the transfer functions are matrices. Let us first recall what we have already done. The transfer matrix $\mathbf{H}_N(z)$ from the prediction error $\mathbf{e}_N^f(n)$ to the prediction error $\mathbf{e}_N^b(n)$ is given in Eq. (8.48) and is reproduced below:

$$z^{-1}\mathbf{H}_N(z) \stackrel{\Delta}{=} \mathbf{B}_N(z)\mathbf{A}_N^{-1}(z). \qquad (8.89)$$

The transfer matrix $z^{-1}\mathbf{G}_N(z)$ from the *normalized* prediction error $\mathbf{g}_N^f(n)$ to the normalized prediction error $\mathbf{g}_N^b(n)$ is given in Eq. (8.50) and reproduced below:

$$z^{-1}\mathbf{G}_N(z) \stackrel{\Delta}{=} [\mathbf{E}_N^b]^{-1/2}\mathbf{B}_N(z)\mathbf{A}_N^{-1}(z)[\mathbf{E}_N^f]^{1/2}. \qquad (8.90)$$

This also appears in the MIMO IIR lattice of Fig. 8.8(a) (indicated also in Fig. 8.11). We showed in Section 8.11.2 that $\mathbf{G}_N(z)$ is paraunitary, which is an extension of the all-pass property to the MIMO case. We also showed (Section 8.11.3) that all poles of $\mathbf{G}_N(z)$ are in $|z| < 1$.

We will start from these results and establish some more. The fact that the poles of $\mathbf{G}_N(z)$ are in $|z| < 1$ does not automatically imply that the zeros of the determinant of $\mathbf{A}_N(z)$ are in $|z| < 1$ unless we establish that the rational form in Eq. (8.90) is "irreducible" or "minimal" in some sense. We now discuss this issue more systematically. We will also establish a relation between the forward and backward matrix polynomials $\mathbf{A}_N(z)$ and $\mathbf{B}_N(z)$.

8.13.1   Review of Matrix Fraction Descripions

An expression of the form

$$\mathbf{H}(z) = \mathbf{B}(z)\mathbf{A}^{-1}(z), \qquad (8.91)$$

where $\mathbf{A}(z)$ and $\mathbf{B}(z)$ are polynomial matrices is called a matrix fraction description (MFD).[5] The expressions (8.89) and (8.90) for $z^{-1}\mathbf{H}_N(z)$ and $z^{-1}\mathbf{G}_N(z)$ are therefore MFDs. We know that the

---

[5]In this section, all polynomials are polynomials in z^{-1}.

rational expression $H(z) = B(z)/A(z)$ for a transfer function is not unique because there can be cancellations between the polynomials $A(z)$ and $B(z)$. Similarly, the MFD expression for a rational transfer matrix is not unique. If there are "common factors" between $\mathbf{A}(z)$ and $\mathbf{B}(z)$, then we can cancel these to obtain a reduced MFD description for the same $\mathbf{H}(z)$. To be more quantitative, we first need to define "common factors" for matrix polynomials. If the two polynomials $\mathbf{A}(z)$ and $\mathbf{B}(z)$ can be written as

$$\mathbf{B}(z) = \mathbf{B}_1(z)\mathbf{R}(z) \quad \text{and} \quad \mathbf{A}(z) = \mathbf{A}_1(z)\mathbf{R}(z)$$

where $\mathbf{B}_1(z), \mathbf{A}_1(z)$, and $\mathbf{R}(z)$ are polynomial matrices, then we say that $\mathbf{R}(z)$ is a right common divisor or factor between $\mathbf{A}(z)$ and $\mathbf{B}(z)$, abbreviated as **rcd**. In this case, we can rewrite Eq. (8.91) as

$$\mathbf{H}(z) = \mathbf{B}_1(z)\mathbf{A}_1^{-1}(z). \tag{8.92}$$

Because

$$\det \mathbf{A}(z) = \det \mathbf{A}_1(z)\det \mathbf{R}(z),$$

it follows that $\det \mathbf{A}_1(z)$ has smaller degree than $\det \mathbf{A}(z)$ (unless $\det \mathbf{R}(z)$ is constant). Thus, cancellation of all these common factors results in a reduced MFD (i.e., one with smaller degree for $\det \mathbf{A}(z)$). We say that the MFD (8.91) is *irreducible* if every rcd $\mathbf{R}(z)$ has unit determinant,

$$\det \mathbf{R}(z) = 1, \tag{8.93}$$

or more generally, a nonzero constant. A matrix polynomial $\mathbf{R}(z)$ satisfying Eq. (8.93) is said to be unimodular. Thus, an MFD (8.91) is irreducible if the matrices $\mathbf{A}(z)$ and $\mathbf{B}(z)$ can have only *unimodular* rcds. In this case, we also say that $\mathbf{A}(z)$ and $\mathbf{B}(z)$ are *right coprime*.

Any unimodular matrix is an rcd. Given $\mathbf{A}(z)$ and $\mathbf{B}(z)$, we can always write

$$\mathbf{A}(z) = \underbrace{\mathbf{A}(z)\mathbf{U}^{-1}(z)}_{\text{polynomial } \mathbf{A}_1(z)} \times \mathbf{U}(z), \quad \mathbf{B}(z) = \underbrace{\mathbf{B}(z)\mathbf{U}^{-1}(z)}_{\text{polynomial } \mathbf{B}_1(z)} \times \mathbf{U}(z)$$

for any unimodular polynomial $\mathbf{U}(z)$. The inverse $\mathbf{U}^{-1}(z)$ remains a polynomial because $\det \mathbf{U}(z) = 1$. Thus, $\mathbf{A}_1(z)$ and $\mathbf{B}_1(z)$ are still polynomials. This shows that any unimodular matrix can be regarded as an rcd of any pair of matrix polynomials. □

The following results are well-known in linear system theory (e.g., see Lemmas 13.5.2 and 13.6.1 in Vaidyanathan, 1993).

Lemma 8.4. *Irreducible MFDs and Poles.* Let $\mathbf{H}(z) = \mathbf{B}(z)\mathbf{A}^{-1}(z)$ be an irreducible MFD. Then, the set of poles of $\mathbf{H}(z)$ is precisely the set of zeros of $[\det \mathbf{A}(z)]$.   ◇

Lemma 8.5. *Irreducible and Reducible MFDs.* Let $\mathbf{H}(z) = \mathbf{B}(z)\mathbf{A}^{-1}(z)$ be an MFD and $\mathbf{H}(z) = \mathbf{B}_{\mathrm{red}}(z)\mathbf{A}_{\mathrm{red}}^{-1}(z)$ an irreducible MFD for the same system $\mathbf{H}(z)$. Then, $\mathbf{B}(z) = \mathbf{B}_{\mathrm{red}}(z)\mathbf{R}(z)$ and $\mathbf{A}(z) = \mathbf{A}_{\mathrm{red}}(z)\mathbf{R}(z)$ for some polynomial matrix $\mathbf{R}(z)$. $\diamondsuit$

8.13.2   Relation Between Predictor Polynomials

We now prove a crucial property of the optimal MIMO predictor polynomials. The property depends on the vector version of Levinson's upward recursion Eqs. (8.37(f))–(8.37(g)). First note that we can solve for $\mathbf{A}_m(z)$ and $\mathbf{B}_m(z)$ in terms of $\mathbf{A}_{m+1}(z)$ and $\mathbf{B}_{m+1}(z)$ and obtain

$$\mathbf{A}_m(z) = \left(\mathbf{I} - \mathbf{K}_{m+1}^f(\mathbf{K}_{m+1}^b)^\dagger\right)^{-1}\left(\mathbf{A}_{m+1}(z) - z\mathbf{K}_{m+1}^f\mathbf{B}_{m+1}(z)\right) \qquad (8.94)$$

and

$$\mathbf{B}_m(z) = \left(\mathbf{I} - (\mathbf{K}_{m+1}^b)^\dagger\mathbf{K}_{m+1}^f\right)^{-1}\left(z\mathbf{B}_{m+1}(z) - (\mathbf{K}_{m+1}^b)^\dagger\mathbf{A}_{m+1}(z)\right). \qquad (8.95)$$

This is called the *downward recursion* because we go from $m + 1$ to m using these equations. The inverses indicated in Eq. (8.94) and (8.95) exist by virtue of the assumption that the error covariances at stage m and $m + 1$ are nonsingular and related as in Eqs. (8.35) and (8.36).

Lemma 8.6. *Coprimality of Predictor Polynomials.* The optimal predictor polynomials $\mathbf{A}_N(z)$ and $\mathbf{B}_N(z)$ are right coprime for any N, so that the MFD given by $\mathbf{B}_N(z)\mathbf{A}_N^{-1}(z)$ is irreducible. $\diamondsuit$

Proof. If $\mathbf{R}(z)$ is an rcd of $\mathbf{A}_{m+1}(z)$ and $\mathbf{B}_{m+1}(z)$, then we see from Eqs. (8.94) and (8.95) that it is also an rcd of $\mathbf{A}_m(z)$ and $\mathbf{B}_m(z)$. Repeating this argument, we see that any rcd $\mathbf{R}(z)$ of $\mathbf{A}_N(z)$ and $\mathbf{B}_N(z)$ for $N > 0$ must be an rcd of $\mathbf{A}_0(z)$ and $\mathbf{B}_0(z)$. But $\mathbf{A}_0(z) = \mathbf{I}$ and $\mathbf{B}_0(z) = z^{-1}\mathbf{I}$, so these have no rcd other than unimodular matrices. Thus, $\mathbf{R}(z)$ has to be unimodular, or equivalently, $\mathbf{A}_N(z)$ and $\mathbf{B}_N(z)$ are right coprime. $\square$

We are now ready to prove the following result. The only assumption is that the error covariances $\mathbf{E}_N^f$ and $\mathbf{E}_N^b$ are nonsingular (as assumed throughout the chapter).

Theorem 8.3. *Minimum-Phase Property.* The optimal predictor polynomial matrix $\mathbf{A}_N(z)$ is a minimum-phase polynomial, that is, all the zeros of $[\det\mathbf{A}_N(z)]$ are in $|z| < 1$. $\diamondsuit$

Proof. From Eq. (8.90), we have

$$\mathbf{B}_N(z)\mathbf{A}_N^{-1}(z) = z^{-1}(\mathbf{E}_N^b)^{1/2}\mathbf{G}_N(z)(\mathbf{E}_N^f)^{-1/2}.$$

From Theorem 8.2, we know that $\mathbf{G}_N(z)$ is a stable transfer matrix (all poles restricted to $|z| < 1$). This shows that $\mathbf{B}_N(z)\mathbf{A}_N^{-1}(z)$ is stable as well. Because $\mathbf{B}_N(z)\mathbf{A}_N^{-1}(z)$ is an irreducible MFD (Lemma 8.6), all the zeros of $[\det\mathbf{A}_N(z)]$ are poles of $\mathbf{B}_N(z)\mathbf{A}_N^{-1}(z)$ (Lemma 8.4). So all zeros of $[\det\mathbf{A}_N(z)]$ are in $|z| < 1$. $\square$

Refer again to Eq. (8.90), which is reproduced below:

$$z^{-1}\mathbf{G}_N(z) \triangleq \underbrace{[\mathbf{E}_N^b]^{-1/2}\mathbf{B}_N(z)}_{\text{call this } \mathbf{B}(z)} \underbrace{\mathbf{A}_N^{-1}(z)[\mathbf{E}_N^f]^{1/2}}_{\text{call this } \mathbf{A}^{-1}(z)} = \mathbf{B}(z)\mathbf{A}^{-1}(z) \qquad (8.96)$$

Because $\mathbf{G}_N(z)$ is paraunitary, it follows that $\mathbf{B}(z)\mathbf{A}^{-1}(z)$ is paraunitary as well. Thus,

$$\widetilde{\mathbf{A}}^{-1}(z)\widetilde{\mathbf{B}}(z)\mathbf{B}(z)\mathbf{A}^{-1}(z) = \mathbf{I}$$

from which it follows that

$$\widetilde{\mathbf{B}}(z)\mathbf{B}(z) = \widetilde{\mathbf{A}}(z)\mathbf{A}(z) \quad \text{for all } z. \qquad (8.97)$$

That is,

$$\widetilde{\mathbf{B}}_N(z)[\mathbf{E}_N^b]^{-1}\mathbf{B}_N(z) = \widetilde{\mathbf{A}}_N(z)[\mathbf{F}_N^f]^{-1}\mathbf{A}_N(z). \qquad (8.98)$$

So, the forward and backward predictor polynomials are related by the above equation. For the scalar case, note that we had $\mathcal{E}_N^b = \mathcal{E}_N^f$, so that this reduces to $\widetilde{B}_N(z)B_N(z) = \widetilde{A}_N(z)A_N(z)$, which, of course, is equivalent to the statement that $B_N(z)/A_N(z)$ is all-pass. Summarizing, the main properties of the optimal predictor polynomials for vector LPC are the following:

1. $\mathbf{A}_N(z)$ and $\mathbf{B}_N(z)$ are right coprime, that is, $\mathbf{B}_N(z)\mathbf{A}_N^{-1}(z)$ is irreducible (Lemma 8.6).
2. $\mathbf{A}_N(z)$ has minimum-phase property (Theorem 8.3).
3. $\widetilde{\mathbf{B}}_N(z)[\mathbf{E}_N^b]^{-1}\mathbf{B}_N(z) = \widetilde{\mathbf{A}}_N(z)[\mathbf{E}_N^f]^{-1}\mathbf{A}_N(z)$, or equivalently,

$$z^{-1}\mathbf{G}_N(z) \triangleq [\mathbf{E}_N^b]^{-1/2}\mathbf{B}_N(z)\mathbf{A}_N^{-1}(z)[\mathbf{E}_N^f]^{1/2} \qquad (8.99)$$

is paraunitary.

8.14 CONCLUDING REMARKS

Although the vector LPC theory proceeds along the same lines as does the scalar LPC theory, there are a number of important differences as shown in this chapter. The scalar all-pass lattice structure introduced in Section 4.3 has applications in the design of robust digital filter structures (Gray and Markel, 1973, 1975, 1980; Vaidyanathan and Mitra, 1984, 1985; Vaidyanathan et al., 1986). Similarly, the paraunitary lattice structure for MIMO linear prediction can be used for the design of low sensitivity digital filter structures, as shown in the pioneering work of Rao and Kailath (1984). The MIMO lattice is also studied by Vaidyanathan and Mitra (1985).

· · · ·

APPENDIX A

Linear Estimation of Random Variables

Let $x_1, x_2, \ldots, x_N$ be a set of N random variables, possibly complex. Let x be another random variable, possibly correlated to x_i in some way. We wish to obtain an estimate of x from the observed values of x_i, using a linear combination of the form

$$\widehat{x} = -a_1^* x_1 - a_2^* x_2 \ldots - a_N^* x_N. \qquad (A.1)$$

The use of $-a_i^*$ rather than just a_i might appear to be an unnecessary complication, but will become notationally convenient. We say that $\widehat{x}$ is a *linear estimate* of x in terms of the "observed" variables x_i. The estimation error is defined to be

$$e = x - \widehat{x}, \qquad (A.2)$$

so that

$$e = x + a_1^* x_1 + a_2^* x_2 + \ldots + a_N^* x_N. \qquad (A.3)$$

A classic problem in estimation theory is the identification of the constants a_i such that the mean square error

$$\mathcal{E} = E[\,|e|^2\,], \qquad (A.4)$$

is minimized. The estimate $\widehat{x}$ is then said to be the minimum mean square error (MMSE) linear estimate.

A.1 THE ORTHOGONALITY PRINCIPLE

The minimization of mean square error is achieved with the help of the following fundamental result:

Theorem A.1. *Orthogonality Principle.* The estimate $\widehat{x}$ in Eq. (A.1) results in minimum mean square error among all linear estimates, if and only if the estimation error e is orthogonal to the N random variables x_i, that is, $E[ex_i^*] = 0$ for $1 \leq i \leq N$. $\qquad \diamond$

Proof. Let $\widehat{x}_\perp$ denote the estimate obtained when the orthogonality condition is satisfied. Thus, the error

$$e_\perp \stackrel{\Delta}{=} x - \widehat{x}_\perp \qquad (A.5)$$

satisfies

$$E[e_\perp x_i^*] = 0, \quad 1 \le i \le N. \qquad (A.6)$$

Let $\widehat{x}$ be some other linear estimate, with error e. Then,

$$e = x - \widehat{x} = e_\perp + \widehat{x}_\perp - \widehat{x} \qquad \text{(from Eq. (A.5))}.$$

This has the mean square value

$$E[\,|e|^2] = E[\,|e_\perp|^2] + E[\,|\widehat{x}_\perp - \widehat{x}|^2] + E[(\widehat{x}_\perp - \widehat{x})e_\perp^*] + E[e_\perp(\widehat{x}_\perp^* - \widehat{x}^*)]. \qquad (A.7)$$

Because $\widehat{x}_\perp$ and $\widehat{x}$ are both linear combinations as in Eq. (A.1), we can write

$$\widehat{x}_\perp - \widehat{x} = \sum_{i=1}^{N} c_i x_i. \qquad (A.8)$$

We therefore have

$$E[e_\perp(\widehat{x}_\perp^* - \widehat{x}^*)] = E[e_\perp \sum_{i=1}^{N} c_i^* x_i^*] \quad \text{(from Eq. (A.8))}$$

$$= \sum_{i=1}^{N} c_i^* E[e_\perp x_i^*]$$

$$= 0 \qquad \text{(from Eq. (A.6))},$$

so that Eq. (A.7) becomes

$$E[\,|e|^2] = E[\,|e_\perp|^2] + E[\,|\widehat{x}_\perp - \widehat{x}|^2].$$

The second term on the right-hand side above is nonnegative, and we conclude that

$$E[\,|e|^2] \ge E[|e_\perp|^2].$$

Equality holds in the above, if and only if $\widehat{x}_\perp = \widehat{x}$. $\qquad\qquad \square$

A.2   CLOSED-FORM SOLUTION

If we impose the orthogonality condition on the error (A.3), we obtain the N equations

$$E[ex_i^*] = 0, \quad 1 \le i \le N \quad \text{(orthogonality)}. \qquad (A.9)$$

We use these to find the coefficients a_i, which result in the optimal estimate. (The subscript $\perp$ which was used in the above proof will be discontinued for simplicity.) Defining the column vectors

$$\mathbf{x} = \begin{bmatrix} x_1 & x_2 & \ldots & x_N \end{bmatrix}^{\mathrm{T}}, \quad \mathbf{a} = \begin{bmatrix} a_1 & a_2 & \ldots & a_N \end{bmatrix}^{\mathrm{T}},$$

we can write the error e as

$$e = x + \mathbf{a}^{\dagger}\mathbf{x}, \tag{A.10}$$

and the orthogonality condition Eq. (A.9) as

$$E[e\mathbf{x}^{\dagger}] = \mathbf{0}. \tag{A.11}$$

Substituting from Eq. (A.10), we arrive at

$$E[x\mathbf{x}^{\dagger}] + \mathbf{a}^{\dagger}E[\mathbf{x}\mathbf{x}^{\dagger}] = \mathbf{0}. \tag{A.12}$$

In the above equation, the matrix

$$\mathbf{R} \overset{\Delta}{=} E[\mathbf{x}\mathbf{x}^{\dagger}]$$

is the correlation matrix for the random vector $\mathbf{x}$. This matrix is Hermitian, that is, $\mathbf{R}^{\dagger} = \mathbf{R}$. Defining the vector

$$\mathbf{r} = E[\mathbf{x}\,x^*],$$

Eq. (A.12) simplifies to

$$\mathbf{R}\mathbf{a} = -\mathbf{r}.$$

More explicitly, in terms of the matrix elements, this can be written as

$$\underbrace{\begin{bmatrix} E[\,|x_1|^2] & E[x_1 x_2^*] & \ldots & E[x_1 x_N^*] \\ E[x_2 x_1^*] & E[\,|x_2|^2] & \ldots & E[x_2 x_N^*] \\ \vdots & \vdots & \ddots & \vdots \\ E[x_N x_1^*] & E[x_N x_2^*] & \ldots & E[\,|x_N|^2] \end{bmatrix}}_{\mathbf{R}} \underbrace{\begin{bmatrix} a_1 \\ a_2 \\ \vdots \\ a_N \end{bmatrix}}_{\mathbf{a}} = -\underbrace{\begin{bmatrix} E[x_1 x^*] \\ E[x_2 x^*] \\ \vdots \\ E[x_N x^*] \end{bmatrix}}_{-\mathbf{r}}. \tag{A.13}$$

The vector $\mathbf{r}$ is the cross-correlation vector. Its ith element represents the correlation between the random variable x and the observations x_i. If the correlation matrix $\mathbf{R}$ is nonsingular, we can solve for $\mathbf{a}$ and obtain

$$\mathbf{a} = -\mathbf{R}^{-1}\mathbf{r} \quad \text{(optimal linear estimator)}. \tag{A.14}$$

The autocorrelation matrix $\mathbf{R}$ is positive semidefinite because $\mathbf{x}\mathbf{x}^\dagger$ is positive semidefinite. If in addition $\mathbf{R}$ is nonsingular, then it is positive definite. The case of singular $\mathbf{R}$ will be discussed in Section A.4.

The above development does not assume that x or x_i have zero mean. Notice that if the correlation between x and the observation x_i is zero for each i, then $\mathbf{r} = 0$. Under this condition, $\mathbf{a} = \mathbf{0}$. This means that the best estimate is $\widehat{x} = 0$ and the error is $e = x$ itself!

A.3 CONSEQUENCES OF ORTHOGONALITY

Because the estimate $\widehat{x}$ is a linear combination of the random variables x_i, the orthogonality property (A.9) implies

$$E[e\widehat{x}^*] = 0. \qquad (A.15)$$

That is, the optimal linear estimate $\widehat{x}$ is itself orthogonal to the error e. The random variable x and its estimate $\widehat{x}$ are related as $x = \widehat{x} + e$. From this, we obtain

$$E[\,|x|^2] = E[\,|\widehat{x}|^2] + \underbrace{E[\,|e|^2]}_{\mathcal{E}} \qquad (A.16)$$

by exploiting Eq. (A.15).

Geometric Interpretation. A geometric interpretation of orthogonality principle can be given by visualizing the random variables as vectors. Imagine we have a three-dimensional Euclidean vector $\mathbf{x}$, which should be approximated in the two-dimensional plane (x_1, x_2) (Fig. A.1). We see that the approximation error is given by the vector $\mathbf{e}$ and that the length of this vector is minimized if and only if it is perpendicular to the x_1, x_2 plane. Under this condition, the lengths of the vector $\mathbf{x}$, the error $\mathbf{e}$, and the estimate $\widehat{x}$ are indeed related as

$$(\text{length of } \mathbf{x})^2 = (\text{length of } \widehat{\mathbf{x}})^2 + (\text{length of } \mathbf{e})^2,$$

which is similar to Eq. (A.16). The above relation is also reminiscent of the Pythagorean theorem. □

Expression for the Minimized Error. Assuming that the coefficients a_i have been found, we can compute the minimized mean square error as follows:

$$\mathcal{E} = E[|e|^2] = E[ee^*] = E[e(x^* + \mathbf{x}^\dagger\mathbf{a})]$$
$$= E[ex^*] \quad \text{(by orthogonality (A.11))}.$$

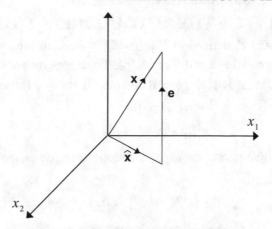

FIGURE A.1: Pertains to the discussion of orthogonality principle.

Using Eq. (A.10) and the definition of **r**, this can be rewritten as

$$\underbrace{E[\,|e|^2\,]}_{\mathcal{E}} = E[\,|x|^2\,] + \mathbf{a}^\dagger \mathbf{r} \qquad\qquad (A.17)$$

Example A.1: Estimating a Random Variable. Let x and x_1 be real random variables. We wish to estimate x from the observed values of x_1 using the linear estimate

$$\widehat{x} = -a_1 x_1.$$

The normal equations are obtained by setting $N = 1$ in Eq. (A.13), that is,

$$E[x_1^2] \times a_1 = -E[x_1 x].$$

Thus,

$$a_1 = \frac{-E[x_1 x]}{E[x_1^2]}.$$

The mean square value of the estimation error is

$$E[e^2] = E[x^2] - \frac{(E[x_1 x])^2}{E[x_1^2]} \quad \text{[from Eq. (A.17)]}.$$

Thus, a_1 is proportional to the cross-correlation $E[x_1 x]$. If $E[x_1 x] = 0$, then $a_1 = 0$, that is, the best estimate is $\widehat{x} = 0$, and the mean square error is

$$E[e^2] = E[x^2] = \quad \text{mean square value of } x \text{ itself!}$$

A.4 SINGULARITY OF THE AUTOCORRELATION MATRIX

If the autocorrelation matrix $\mathbf{R}$ defined in Eq. (A.13) is singular, (i.e., $[\det \mathbf{R}] = 0$), we cannot invert it to obtain a unique solution $\mathbf{a}$ as in Eq. (A.14). To analyze the meaning of singularity, recall (Horn and Johnson, 1985) that $\mathbf{R}$ is singular if and only if $\mathbf{Rv} = 0$ for some vector $\mathbf{v} \neq \mathbf{0}$. Thus, singularity of $\mathbf{R}$ implies $\mathbf{v}^\dagger \mathbf{Rv} = 0$, that is,

$$\mathbf{v}^\dagger E[\mathbf{xx}^\dagger]\mathbf{v} = 0, \quad \mathbf{v} \neq \mathbf{0}.$$

Because $\mathbf{v}$ is a constant with respect to the expectation operation, we can rewrite this as $E[\mathbf{v}^\dagger \mathbf{xx}^\dagger \mathbf{v}] = 0$. That is,

$$E[\,|\mathbf{v}^\dagger \mathbf{x}|^2] = 0.$$

This implies that the scalar random variable $\mathbf{v}^\dagger \mathbf{x}$ is zero, that is,

$$\mathbf{v}^\dagger \mathbf{x} = v_1^* x_1 + v_2^* x_2 + \ldots + v_N^* x_N = 0.$$

with $\mathbf{x} = \begin{bmatrix} x_1 & x_2 & \ldots & x_N \end{bmatrix}^{\mathrm{T}}$ and $\mathbf{v} = \begin{bmatrix} v_1 & v_2 & \ldots & v_N \end{bmatrix}^{\mathrm{T}}$. Because $\mathbf{v} \neq \mathbf{0}$, at least one of components v_i is nonzero. In other words, the N random variables x_i are *linearly dependent*. Assuming, for example, that v_N is nonzero, we can write

$$x_N = -[v_1^* x_1 + \ldots + v_{N-1}^* x_{N-1}]/v_N^*.$$

That is, we can drop the term $a_N^* x_N$ from Eq. (A.3) and solve a smaller estimation problem, without changing the value of the optimal estimate. We can continue to eliminate the linear dependence among random variables in this manner until we finally arrive at a smaller set of random variables, say, x_i, $1 \leq i \leq L$, which do not have linear dependence. The $L \times L$ correlation matrix of these random variables is then nonsingular, and we can solve a set of L equations (normal equations) similar to Eq. (A.13), to find the unique set of L coefficients a_i.

Example A.2: Singularity of Autocorrelation. Consider the estimation of x from two real random variables x_1 and x_2. Let the 2×2 autocorrelation matrix be

$$\mathbf{R} = \begin{bmatrix} 1 & a \\ a & a^2 \end{bmatrix}.$$

We have $[\det \mathbf{R}] = a^2 - a^2 = 0$ so that $\mathbf{R}$ is singular. We see that the vector $\mathbf{v} = \begin{bmatrix} a \\ -1 \end{bmatrix}$ has the property that $\mathbf{Rv} = \mathbf{0}$. From the above theory, we conclude that x_1 and x_2 are linearly dependent and $ax_1 - x_2 = 0$, that is,

$$x_2 = ax_1.$$

In other words, x_2 is completely determined by x_1.

APPENDIX B

Proof of a Property of Autocorrelations

The autocorrelation $R(k)$ of a WSS process satisfies the inequality

$$R(0) \geq |R(k)|. \tag{B.1}$$

To prove this, first observe that

$$E|x(0) - e^{j\phi}x(-k)|^2 \geq 0 \tag{B.2}$$

for any real constant ϕ. The left-hand side can be rewritten as

$$E\left[|x(0)|^2 + |x(-k)|^2 - e^{j\phi}x^*(0)x(-k) - e^{-j\phi}x(0)x^*(-k)\right]$$
$$= 2R(0) - e^{j\phi}R^*(k) - e^{-j\phi}R(k).$$

Because this holds for any real ϕ, let us choose ϕ so that $R^*(k)e^{j\phi}$ is real and nonnegative, that is, $R^*(k)e^{j\phi} = |R(k)|$. Thus,

$$E|x(0) - e^{j\phi}x(-k)|^2 = 2R(0) - 2|R(k)| \geq 0,$$

from (B.2). This proves Eq. (B.1). Equality in (B.1) is possible if and only if

$$E|x(0) - e^{j\phi}x(-k)|^2 = 0,$$

that is,

$$E|x(n) - e^{j\phi}x(n - k)|^2 = 0$$

for all n (by WSS property). Thus,

$$x(n) = e^{j\phi}x(n - k),$$

which means that all samples of $x(n)$ can be predicted perfectly if we know the first k samples $x(0), \ldots, x(k-1)$. Summarizing, as long as the process is not perfectly predictable, we have

$$R(0) > |R(k)| \tag{B.3}$$

for $k \neq 0$.

APPENDIX C

Stability of the Inverse Filter

We now give a direct proof that the optimum prediction polynomial $A_N(z)$ has all zeros p_k inside the unit circle, that is, $|p_k| < 1$, as long as $x(n)$ is not a line spectral process (i.e., not fully predictable). This gives a direct proof (without appealing to Levinson's recursion) that the causal IIR filter $1/A_N(z)$ is stable. This proof appeared in Vaidyanathan et al. (1997). Also see Lang and McClellan (1979).

Proof. The prediction filter reproduced in Fig. C.1(a) can be redrawn as in Fig. C.1(b), where q is some zero of $A_N(z)$ and $C_{N-1}(z)$ is causal FIR with order $N-1$. First observe that $1 - qz^{-1}$ is the optimum first-order prediction polynomial for the WSS process $y(n)$, for otherwise, the mean square value of its output can be made smaller by using a different q, which contradicts the fact that $A_N(z)$ is optimal. Thus, q is the optimal coefficient

$$ q = \frac{R_{yy}(1)}{R_{yy}(0)}, \qquad (C.1) $$

where $R_{yy}(k)$ is the autocorrelation of the WSS process $y(n)$. From Appendix B, we know that $R_{yy}(0) \geq |R_{yy}(k)|$ for any k that shows

$$ |q| \leq 1. $$

This proves that all zeros of $A_N(z)$ satisy $|z_k| \leq 1$. To show that, in fact, $|q| < 1$ when $x(n)$ is not a line spectral process, we have to work a little harder. Recall that the optimal $e_N^f(n)$ is orthogonal to $x(n-1), \ldots, x(n-N)$ (Section 2.3). Because $y(n-1)$ is a linear combination of $x(n-1), \ldots, x(n-N)$, it is orthogonal to $e_N^f(n)$ as well:

$$ E[e_N^f(n)y^*(n-1)] = 0. \qquad (C.2) $$

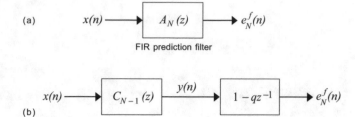

FIGURE C.1: (a) The prediction filter and (b) an equivalent drawing, with a factor $(1 - qz^{-1})$ shown separately.

Thus, using the fact that $e_N^f(n) = y(n) - qy(n-1)$, we can write

$$
\begin{aligned}
\mathcal{E}_N^f = E[|e_N^f(n)|^2] &= E[e_N^f(n)(y(n) - qy(n-1))^*] \\
&= E[e_N^f(n)y^*(n)] \quad \text{(from (C.2))} \\
&= E[(y(n) - qy(n-1))y^*(n)] \\
&= R_{yy}(0) - qR_{yy}^*(1) \\
&= R_{yy}(0)(1 - |q|^2) \quad \text{(from (C.1))}.
\end{aligned}
$$

Because $x(n)$ is not a line spectral process, we have $\mathcal{E}_N^f \neq 0$, that is, $\mathcal{E}_N^f > 0$. Thus, $(1 - |q|^2) > 0$, which proves $|q| < 1$ indeed. $\square$

•   •   •   •

APPENDIX D

Recursion Satisfied by AR Autocorrelations

Let $a_{k,n}^f$ denote the optimal Nth-order prediction coefficients and $\mathcal{E}_N^f$ the corresponding prediction error for a WSS process $x(n)$ with autocorrelation $R(k)$. Then, the autocorrelation satisfies the following recursive equations:

$$R(k) = \begin{cases} -a_{N,1}^* R(k-1) - a_{N,2}^* R(k-2) \ldots - a_{N,N}^* R(k-N) + \mathcal{E}_N^f, & k = 0 \\ -a_{N,1}^* R(k-1) - a_{N,2}^* R(k-2) \ldots - a_{N,N}^* R(k-N), & 1 \le k \le N. \end{cases} \qquad (D.1)$$

In particular, if the process $x(n)$ is AR(N), then the second equation can be made stronger:

$$R(k) = -a_{N,1}^* R(k-1) - a_{N,2}^* R(k-2) \ldots - a_{N,N}^* R(k-N), \quad \text{for all } k > 0 \qquad (D.2)$$

Proof. The above equations are nothing but the augmented normal equations rewritten. To see this, note that the 0th row in Eq. (2.20) is precisely the $k = 0$ case in (D.1). For $1 \le k \le N$, the kth equation in (2.20) is

$$R^*(k) + a_{N,1} R^*(k-1) + \ldots + a_{N,N} R^*(k-N) = 0 \text{ (using } R(i) = R^*(-i)).$$

Conjugating both sides, we get the second equation in Eq. (D.1). For the case of the AR(N) process, we know that the $(N+m)$th-order optimal predictor for any $m > 0$ is the same as the Nth-order optimal predictor, that is,

$$a_{N+m,n} = \begin{cases} a_{N,n} & 1 \le n \le N \\ 0 & n > N. \end{cases}$$

So, the second equation in Eq. (D.1), written for $N + m$ instead of N yields

$$R(k) = -a_{N,1}^* R(k-1) - a_{N,2}^* R(k-2) \ldots - a_{N,N}^* R(k-N), \ 1 \le k \le N+m$$

for any $m \ge 0$. This proves Eq. (D.2) for AR(N) processes. $\qquad \square$

$\bullet \quad \bullet \quad \bullet \quad \bullet$

Problems

1. Suppose $x(n)$ is a zero-mean white WSS random process with variance σ^2. What are the coefficients $a_{N,i}$ of the Nth-order optimal predictor? What is the prediction error variance?

2. Consider the linear prediction problem for a real WSS process $x(n)$ with autocorrelation $R(k)$. The prediction coefficients $a_{N,i}$ are now real. Show that the mean square prediction error $\mathcal{E}_N^f$ can be written as

$$\mathcal{E}_N^f = R(0) + \mathbf{a}^\mathrm{T} \mathbf{R}_N \mathbf{a} + 2\mathbf{a}^\mathrm{T} \mathbf{r},$$

where

$$\mathbf{a} = \left[a_{N,1}\ a_{N,2} \cdots a_{N,N} \right]^\mathrm{T},$$

and $\mathbf{R}_N$ and $\mathbf{r}$ are as in Eq. (2.8). By differentiating $\mathcal{E}_N^f$ with respect to each $a_{N,i}$ and setting it to zero, show that we again obtain the normal equations (derived in Section 2.3 using orthogonality principle).

3. Consider a WSS process $x(n)$ with autocorrelation

$$R(k) = \begin{cases} b^{0.5|k|} & \text{for } k \text{ even} \\ 0 & \text{for } k \text{ odd} \end{cases} \tag{P.3}$$

where $0 < b < 1$.
   a) Compute the power spectral density $S_{xx}(e^{j\omega})$.
   b) Find the predictor polynomials $A_1(z), A_2(z)$, the lattice coefficients k_1, k_2, and the error variances $\mathcal{E}_1^f, \mathcal{E}_2^f$ using Levinson's recursion.
   c) Show that $e_2^f(n)$ is white.

4. Someone performs optimal linear prediction on a WSS process $x(n)$ and finds that the first two lattice coefficients are $k_1 = 0.5$ and $k_2 = 0.25$ and that the second-order mean square prediction error is $\mathcal{E}_2^f = 1.0$. Using these values, compute the following.
   a) The predictor polynomials $A_1(z)$ and $A_2(z)$.
   b) The mean square errors $\mathcal{E}_0^f$ and $\mathcal{E}_1^f$.
   c) The autocorrelation coefficients $R(m)$ of the process $x(n)$, for $m = 0, 1, 2$.

5. Someone performs optimal linear prediction on a WSS process $x(n)$ and finds that the first four lattice coefficients are

$$k_1 = 0, \ k_2 = 0.5, \ k_3 = 0, \ k_4 = 0.25, \qquad (P.5)$$

and that the fourth-order mean square prediction error is $\mathcal{E}_4^f = 2.0$. Using these values, compute the following:

a) The predictor polynomials $A_m(z)$ for $m = 1,2,3,4$.
b) The mean square errors $\mathcal{E}_m^f$ for $m = 0,1,2,3$.
c) The autocorrelation coefficients $R(m)$ of the process $x(n)$, for $m = 0,1,2,3,4$.

6. Show that the quantity α_m arising in Levinson's recursion can be expressed in terms of a cross-correlation, that is,

$$\alpha_m^* = E[e_m^f(n)x^*(n - m - 1)]. \qquad (P.6)$$

7. *Estimation From Noisy Data.* Let v be a noisy measurement of a random variable u, that is,

$$v = u + \epsilon, \qquad (P.7a)$$

where the random variable ϵ represents noise . Assume that this noise has zero mean and that it is uncorrelated with u. From this noisy measurement v, we would like to find an estimate of u of the form $\widehat{u} = av$. Let $e = u - \widehat{u}$ be the estimation error. Compute the value of a, which minimizes the mean square error $E[|e|^2]$. Hence, show that the best estimate of u is

$$\widehat{u} = \left(\frac{R_{uu}}{R_{uu} + \sigma_\epsilon^2} \right) v, \qquad (P.7b)$$

where $R_{uu} = E[|u|^2]$ and $\sigma_\epsilon^2 = E[|\epsilon|^2]$ (noise variance). Show also that the minimized mean square error is

$$E[|e|^2] = \frac{R_{uu}\sigma_\epsilon^2}{R_{uu} + \sigma_\epsilon^2}. \qquad (P.7c)$$

Thus, when the noise variance σ_ϵ^2 is very large, the best estimate approaches zero; this means that we should not trust the measurement and just take the estimate to be zero. In this extreme case, the error $E[|e|^2]$ approaches R_{uu}. Thus, with the above estimate, the mean square estimation error will never exceed the mean square value R_{uu} of the random variable u to be estimated, no matter how strong the noise is!

8. *Valid Autocorrelations.* Let **R** be a Hermitian positive definite matrix. Show that it is a valid autocorrelation matrix. That is, show that there exists a random vector **x** such that **R** = $E[\mathbf{x}\mathbf{x}^\dagger]$.

9. Let **J** denote the $N \times N$ *antidiagonal matrix*, demonstrated below for $N = 3$.

$$\mathbf{J} = \begin{bmatrix} 0\ 0\ 1 \\ 0\ 1\ 0 \\ 1\ 0\ 0 \end{bmatrix} \qquad (P.9)$$

If **A** is any $N \times N$ matrix, then **JA** represents a matrix whose rows are identical to those of **A** but renumbered in reverse order. Similarly, **AJ** corresponds to column reversal.

a) Now assume that **A** is Hermitian and Toeplitz. Show that $\mathbf{JAJ} = \mathbf{A}^*$.
b) Conversely, suppose $\mathbf{JAJ} = \mathbf{A}^*$. Does it mean that **A** is Hermitian and Toeplitz? Justify.
c) Suppose $\mathbf{JAJ} = \mathbf{A}^*$ and **A** is Hermitian. Does it mean that **A** is Toeplitz? Justify.

10. Consider a WSS process $x(n)$ with power spectrum

$$S_{xx}(e^{j\omega}) = \frac{1 - \rho^2}{1 + \rho^2 - 2\rho \cos \omega} \qquad -1 < \rho < 1. \qquad (P.10)$$

Let $\mathbf{R}_2$ represent the 2×2 autocorrelation matrix of $x(n)$ as usual. Compute the eigenvalues of this matrix and verify that they are bounded by the extrema of the power spectrum (i.e., as in Eq. (2.28)).

11. *Condition number and spectral dynamic range.* Eq. (2.31) shows that if the power spectrum has wide variations (large S_{max}/S_{min}), then the condition number of $\mathbf{R}_N$ can, in general, be very large. However, this does not necessarily always happen: we can create examples where S_{max}/S_{min} is arbitrarily large, and yet the condition number of $\mathbf{R}_N$ is small. For example, given the autocorrelation $R(k)$ with Fourier transform $S_{xx}(e^{j\omega})$, define a new sequence as follows:

$$R_{yy}(k) = \begin{cases} R(k/N) & \text{if } k \text{ is multiple of } N \\ 0 & \text{otherwise.} \end{cases}$$

a) Find the Fourier transform $S_{yy}(e^{j\omega})$ of $R_{yy}(k)$, in terms of $S_{xx}(e^{j\omega})$.
b) Argue that $R_{yy}(k)$ represents a valid autocorrelation for a WSS process $y(n)$.
c) For the process $y(n)$, write down the $N \times N$ autocorrelation matrix. What is its condition number?

d) For the process $x(n)$, let γ denote the ratio $S_{\max}/S_{\min}$ in its power spectrum $S_{xx}(e^{j\omega})$. For the process $y(n)$, what is the corresponding ratio?

12. Consider two Hermitian matrices $\mathbf{A}$ and $\mathbf{B}$ such that $\mathbf{A}$ is the upper left submatrix of $\mathbf{B}$, that is

$$\mathbf{B} = \begin{bmatrix} \mathbf{A} & \times \\ \times & \times \end{bmatrix}. \qquad (P.12)$$

a) Show that the minimum eigenvalue of $\mathbf{B}$ cannot be larger than that of $\mathbf{A}$. Similarly, show that the maximum eigenvalue of $\mathbf{B}$ cannot be smaller than that of $\mathbf{A}$. *Hint.* Use Rayleigh's principle (Horn and Johnson, 1985).

b) Prove that the condition number $\mathcal{N}$ of the autocorrelation matrix $\mathbf{R}_m$ of a WSS process (Section 2.4.1) cannot decrease as the matrix size m grows.

13. Consider a WSS process $x(n)$ with the first three autocorrelation coefficients

$$R(0) = 3, \; R(1) = 2, \; R(2) = 1. \qquad (P.13)$$

a) Compute the coefficients of the first and the second-order optimal predictor polynomials $A_1(z)$ and $A_2(z)$ by directly solving the normal equations.

b) By using Levinson's recursion, compute the predictor polynomials and mean square prediction errors for the first- and second-order predictors.

c) Draw the second-order FIR LPC lattice structure representing the optimal predictor and indicate the forward and backward prediction errors at the appropriate nodes. Indicate the values of the lattice coefficients k_1 and k_2.

d) Draw the second-order IIR lattice structure representing the optimal predictor and indicate the forward and backward prediction errors at the appropriate nodes. Indicate the values of the lattice coefficients k_1 and k_2.

14. *Valid Toeplitz autocorrelation.* Consider a set of numbers $R(k), 0 \le k \le N$ and define the matrix $\mathbf{R}_{N+1}$ as in Eq. (2.20). By construction, this matrix is Hermitian and Toeplitz. Suppose, in addition, it turns out that it is also positive definite. Then, show that the numbers $R(k), 0 \le k \le N$, are valid autocorrelation coefficients of some WSS process $x(n)$. Give a constructive proof, that is, show how you would generate a WSS process $x(n)$ such that its first $N+1$ autocorrelation coefficients would be the elements $R(0), R(1), \ldots, R(N)$.

15. Show that the coefficients $b_{N,i}$ of the optimal backward predictor are indeed given by reversing and conjugating the coefficients of the forward predictor, as shown in Eq. (4.3).

16. Let $e_m^f(n)$ denote the forward prediction error for the optimal mth-order predictor as usual. Show that the following orthogonality condition is satisfied:

$$E\left[e_m^f(n)[e_k^f(n-1)]^*\right] = 0, \quad m > k. \qquad (P.16)$$

17. In Problem 4, suppose you are informed that $x(n)$ is actually AR(2). Then, compute the autocorrelation $R(m)$ for $m = 3,4$.

18. In Example 2.1, the autocorrelation $R(k)$ of a WSS process $x(n)$ was supplied in closed form. Show that the process $x(n)$ is in fact AR(2), that is, the prediction error $e_2^f(n)$ is white.

19. Compute and sketch the entropy of a Gaussian WSS process, with autocorrelation $R(k) = \rho^{|k|}$, $-1 < \rho < 1$. Make qualitative remarks explaining the nature of the plot.

20. Consider a Gaussian AR(2) process with $R(0) = 1$, for which the second-order predictor polynomial is

$$A_2(z) = \left(1 - \frac{1}{2}z^{-1}\right)\left(1 - \frac{1}{3}z^{-1}\right) \qquad (P.20)$$

Compute the entropy of the process and the flatness measure γ_x^2.

21. Consider a real WSS Gaussian process $x(n)$ with power spectrum sketched as shown in Fig. P.21.
    a) Compute the entropy of the process.

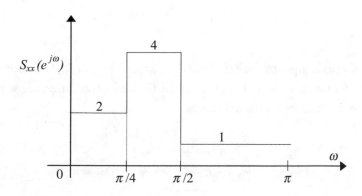

FIGURE P.21

b) Compute the limiting value $\mathcal{E}_\infty^f$ of the mean square prediction error.

c) Compute the flatness measure γ_x^2.

22. Consider a WSS process $x(n)$ with autocorrelation $R(k)$. Let the 3×3 autocorrelation matrix have the form

$$\mathbf{R}_3 = \begin{bmatrix} 1 & \rho & \alpha \\ \rho & 1 & \rho \\ \alpha & \rho & 1 \end{bmatrix}, \tag{P.22}$$

where ρ and α are real and $-1 < \rho < 1$. (Thus, $R(0) = 1$, $R(1) = \rho$, and $R(2) = \alpha$.) If you are told that $x(n)$ is a line spectral process of degree $= 2$, what would be the permissible values of α? For each of these permissible values of α, do the following:

a) Find the line frequencies ω_1 and ω_2 and the powers at these line frequencies.

b) Find the recursive difference equation satisfied by $R(k)$.

c) Find a closed form expression for $R(k)$.

d) Compute $R(3)$ and $R(4)$.

23. We now provide a direct, simpler, proof of the result in Theorem 7.4. Let $y(n)$ be a WSS random process such that if it is input to the FIR filter $1 - \alpha z^{-1}$ the output is identically zero (see Fig. P.23)

a) Show that the constant α *must* have unit magnitude. (Assume, of course, that $y(n)$ itself is not identically zero.)

b) Now let $x(n)$ be a WSS process satisfying the difference equation

$$x(n) = -\sum_{i=1}^{L} a_i^* x(n - i). \tag{P.23}$$

That is, if $x(n)$ is input to the FIR filter $A(z) = 1 + \sum_{i=1}^{L} a_i^* z^{-i}$, the output is identically zero. Furthermore, let there be no such FIR filter of lower order. Show then that all the zeros of $A(z)$ must be on the unit circle.

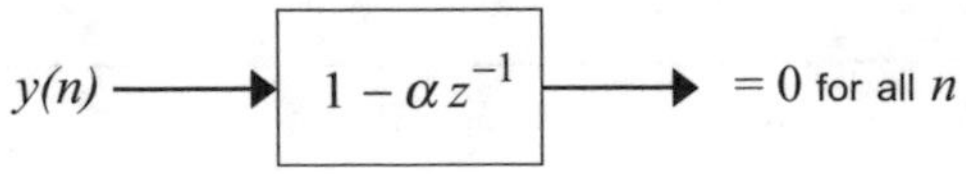

$y(n) \longrightarrow \boxed{1 - \alpha z^{-1}} \longrightarrow = 0 \text{ for all } n$

FIGURE P.23

24. Let $\mathbf{R}_m$ be the $m \times m$ autocorrelation matrix of a Linespec(L) process $x(n)$. We know that $\mathbf{R}_L$ has rank L and $\mathbf{R}_{L+1}$ is singular, with rank L. Using (some or all of) the facts that $\mathbf{R}_m$ is Hermitian, Toeplitz, and positive semidefinite for any m, show the following.
    a) $\mathbf{R}_m$ has rank m for all $m \leq L$.
    b) $\mathbf{R}_m$ has rank exactly L for any $m \geq L$.

25. Show that the sum of sinusoids considered in Example 7.3 is indeed WSS under the three conditions stated in the example.

26. Suppose a sequence of numbers

$$\alpha_0, \alpha_1, \alpha_2, \ldots \qquad (P.26a)$$

tends to a limit α. This means that, given $\epsilon > 0$, there exits a finite integer N, such that

$$|\alpha_i - \alpha| < \epsilon \quad \text{for } i > N. \qquad (P.26b)$$

    a) Show that

$$\lim_{M \to \infty} \frac{1}{M} \sum_{i=0}^{M-1} \alpha_i = \alpha. \qquad (P.26c)$$

    In other words, define a sequence $\beta_M = (1/M) \sum_{i=0}^{M-1} \alpha_i$ and show that it tends to α.
    b) Suppose $f(x)$ is a continuous function for all x in a given range $\mathcal{R}$. Thus, for any fixed x in the range, if we are given a number $\epsilon > 0$, we can find a number $\delta > 0$ such that

$$|f(x_1) - f(x)| < \epsilon \quad \text{for } |x_1 - x| < \delta. \qquad (P.26d)$$

    Let all members α_i of the sequence (P.26a) be in the range $\mathcal{R}$. Show then that

$$\lim_{i \to \infty} f(\alpha_i) = f\left(\lim_{i \to \infty} \alpha_i\right) = f(\alpha). \qquad (P.26e)$$

27. Let $R_1(k)$ and $R_2(k)$ represent the autocorrelations of two WSS random processes. Which of the following sequences represent valid autocorrelations?
    a) $R_1(k) + R_2(k)$
    b) $R_1(k) - R_2(k)$
    c) $R_1(k)R_2(k)$
    d) $\sum_n R_1(n)R_2(k-n)$
    e) $R_1(2k)$

f) $R(k) = \begin{cases} R_1(k/2) & \text{for } k \text{ even} \\ 0 & \text{for } k \text{ odd.} \end{cases}$

28. Let $R(k)$ be the autocorrelation of a zero mean WSS process and suppose $R(k_1) = R(k_2) \neq 0$ for some $k_2 > k_1 > 0$. Does this mean that $R(k)$ is periodic? *Note.* If $k_1 = 0$ and $k_2 > 0$, then the answer is *yes*, as asserted by Theorem 7.2.

29. Find an example of the autocorrelation $R(k)$ for a WSS process such that, for some m, the autocorrelation matrices $\mathbf{R}_m$ and $\mathbf{R}_{m+1}$ have the same minimum eigenvalue, but the minimum eigenvalue of $\mathbf{R}_{m+2}$ is smaller.

30. Let x and y be real random variables such that $E[xy] = E[x]E[y]$, that is, they are uncorrelated. This does not, in general, imply that $E[x^k y^n] = E[x^k]E[y^n]$ (i.e., that x^k and y^n are uncorrelated) for arbitrary positive integers k, n. Prove this statement with an example.

31. It is well-known that the power spectrum $S_{xx}(e^{j\omega})$ of a scalar WSS process is nonnegative (Papoulis, 1965; Peebles, 1987). Using this, show that the power spectral matrix $\mathbf{S_{xx}}(e^{j\omega})$ of a WSS vector process $\mathbf{x}(n)$ is positive semidefinite.

32. Let $x(n)$ and $y(n)$ be zero mean WSS processes, with power spectra $S_{xx}(e^{j\omega})$ and $S_{yy}(e^{j\omega})$. Suppose $S_{xx}(e^{j\omega})S_{yy}(e^{j\omega}) = 0$ for all ω, that is, these spectra do not overlap as demonstrated in the figure.
   a) This does not mean that the processes are necessarily uncorrelated. Demonstrate this.
   b) Assume, however, that the above processes are not only WSS but also jointly WSS (i.e., $E[x(n)y^*(n-k)]$ does not depend on n). Show then that they are indeed uncorrelated.

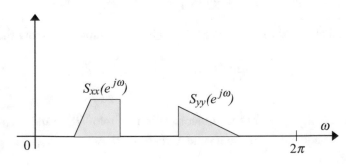

FIGURE P.32

References

1. Anderson, B. D. O., and Moore, J. B. *Optimal filtering*, Prentice-Hall, Englewood Cliffs, NJ, 1979.

2. Antoniou, A. *Digital signal processing: signals, systems, and filters*, McGraw-Hill, New York, 2006.

3. Antoniou, A., and Lu, W.-S. *Practical optimization: algorithms and engineering applications*, Springer, New York, 2007.

4. Atal, B. S., and Schroeder, M. R. "Adaptive predictive coding of speech signals," *Bell. Sys. Tech. J.*, vol. 49, pp. 1973–1986, October 1970. doi:10.1109/ICASSP.1980.1170967

5. Atal, B. S., and Schroeder, M. R. "Predictive coding of speech signals and subjective error criteria," *IEEE Trans. Acoust. Speech Signal Process.*, vol. 27, pp. 247–254, June 1979.

6. Balabanian, N., and Bickart, T. A. *Electrical network theory*, John Wiley & Sons, New York, 1969.

7. Bellman, R. *Introduction to matrix analysis*, McGraw-Hill, New York, 1960.

8. Berlekamp, E. R. *Algebraic coding theory*, McGraw-Hill, New York, 1969.

9. Blahut, R. E. *Fast algorithms for digital signal processing*, Addison-Wesley, Reading, MA, 1985.

10. Burg, J. P. "The relationship between maximum entropy spectra and maximum likelihood spectra," *Geophysics*, vol. 37, pp. 375–376, April 1972. doi:10.1190/1.1440265

11. Chandrasekhar, S. "On the radiative equilibrium of a stellar atmosphere," *Astrophys. J.*, vol. 106, pp. 152–216, 1947.

12. Chellappa, R., and Kashyap, R. L. "Texture synthesis using 2-D noncausal autoregressive models," *IEEE Trans. Acoust. Speech Signal Process.*, vol. 33, pp. 194–203, February 1985.

13. Chong, E. K. P., and Żak, S. H. *An introduction to optimization*, John Wiley & Sons, New York, 2001. doi:10.1109/MAP.1996.500234

14. Churchill, R. V., and Brown, J. W. *Introduction to complex variables and applications*, McGraw-Hill, New York, 1984.

15. Cover, T. M., and Thomas, J. A. *Elements of information theory*, John Wiley & Sons, New York, 1991.

16. Deller, J. R., Proakis, J. G., and Hansen, J. H. L. *Discrete-time processing of speech signals*, Macmillan, New York, 1993.

17. Friedlander, B. "A lattice algorithm for factoring the spectrum of a moving average process," *IEEE Trans. Automat. Control*, vol. 28, pp. 1051–1055, November 1983. doi:10.1109/TAC.1983.1103175

18. Gersho, A., and Gray, R. M. *Vector quantization and signal compression*, Kluwer Academic Press, Boston, 1992.

19. Golub, G. H., and Van Loan, C. F. *Matrix computations*, Johns Hopkins University Press, Baltimore, MD, 1989.

20. Gorokhov, A., and Loubaton, P. "Blind identification of MIMO-FIR systems: a generalized linear prediction approach," *Signal Process.*, vol. 73, pp. 105–124, 1999. doi:10.1016/S0165-1684(98)00187-X

21. Gray, Jr., A. H. "Passive cascaded lattice digital filters," *IEEE Trans. Circuits Syst.*, vol. CAS-27, pp. 337–344, May 1980.

22. Gray, R. M. "On asymptotic eigenvalue distribution of Toeplitz matrices," *IEEE Trans. Inform. Theory*, vol. 18, pp. 725–730, November 1972.

23. Gray, Jr., A. H., and Markel, J. D. "Digital lattice and ladder filter synthesis," *IEEE Trans. Audio Electroacoust.* vol. AU-21, pp. 491–500, December 1973. doi:10.1109/TAU.1973.1162522

24. Gray, Jr., A. H., and Markel, J. D. "A normalized digital filter structure," *IEEE Trans. Acoust. Speech Signal Process.*, vol. ASSP-23, pp. 268–277, June 1975. doi:10.1109/TASSP.1975.1162680

25. Grenander, U., and Szego, G. *Toeplitz forms and their applications*, University of California Press, Berkeley, CA, 1958.

26. Hayes, M. H. *Statistical digital signal processing and modeling*, John Wiley & Sons, New York, 1996.

27. Haykin, S. *Adaptive filter theory*, Prentice-Hall, Upper Saddle River, NJ, 1986 and 2002.

28. Horn R. A., and Johnson, C. R. *Matrix analysis*, Cambridge University Press, Cambridge, 1985.

29. Itakura, F. "Line spectrum representation of linear predictive coefficients of speech signals," *J. Acoust. Soc. Am.*, vol. 57, 1975.

30. Itakura, F., and Saito, S. "A statistical method for estimation of speech spectral density and formant frequencies," *Trans IECE Jpn.*, vol. 53-A, pp. 36–43, 1970.

31. Jayant, N. S., and Noll, P. *Digital coding of waveforms*, Prentice-Hall, Englewood Cliffs, NJ, 1984. doi:10.1016/0165-1684(85)90053-2

32. Kailath, T. "A view of three decades of linear filtering theory," *IEEE Trans. Inform. Theory*, vol. 20, pp. 146–181, March 1974. doi:10.1109/TIT.1974.1055174

33. Kailath, T. *Linear systems*, Prentice Hall, Inc., Englewood Cliffs, NJ, 1980.

34. Kailath, T., Kung S-Y., and Morf, M. "Displacement ranks of a matrix," *Am. Math. Soc.,* vol. 1, pp. 769–773, September 1979. doi:10.1016/0022-247X(79)90124-0

35. Kailath, T., Sayed, A. H., and Hassibi, B. *Linear estimation,* Prentice-Hall, Englewood Cliffs, NJ, 2000.

36. Kang, G. S., and Fransen, L. J. "Application of line-spectrum pairs to low-bit-rate speech coders," *Proc. IEEE Int. Conf. Acoust. Speech Signal Process.,* pp. 7.3.1–7.3.4, May 1995.

37. Kay, S. M., and Marple, S. L., Jr., "Spectrum analysis—a modern perspective," *Proc. IEEE,* vol. 69, pp. 1380–1419, November 1981.

38. Kay, S. M. *Modern spectral estimation: theory and application,* Prentice-Hall, Englewood Cliffs, NJ, 1988.

39. Kolmogorov, A. N. "Interpolation and extrapolation of stationary random sequences," *Izv. Akad. Nauk SSSR Ser. Mat.* 5, pp. 3–14, 1941.

40. Kreyszig, E. *Advanced engineering mathematics,* John Wiley & Sons, New York, 1972.

41. Kumaresan, R., and Tufts, D. W. "Estimating the angles of arrival of multiple plane waves," *IEEE Trans. Aerosp. Electron. Syst.,* vol. 19, pp. 134–139, January 1983.

42. Lang, S. W., and McClellan, J. H. "A simple proof of stability for all-pole linear prediction models," *Proc. IEEE,* vol. 67, pp. 860–861, May 1979.

43. Lepschy, A., Mian, G. A., and Viaro, U. "A note on line spectral frequencies," *IEEE Trans. Acoust. Speech Signal Process.,* vol. 36, pp. 1355–1357, August 1988. doi:10.1109/29.1664

44. Levinson, N. "The Wiener rms error criterion in filter design and prediction," *J. Math. Phys.,* vol. 25, pp. 261–278, January 1947.

45. Lopez-Valcarce, R., and Dasgupta, S. "Blind channel equalization with colored sources based on second-order statistics: a linear prediction approach," *IEEE Trans. Signal Process.,* vol. 49, pp. 2050–2059, September 2001. doi:10.1109/78.942633

46. Makhoul, J. "Linear prediction: a tutorial review," *Proc. IEEE,* vol. 63, pp. 561–580, April 1975.

47. Makhoul, J. "Stable and efficient lattice methods for linear prediction," *IEEE Trans. Acoust. Speech Signal Process.,* vol. 25, pp. 423–428, October 1977.

48. Makhoul, J., Roucos, S., and Gish, H. "Vector quantization in speech coding," *Proc. IEEE,* vol. 73, pp. 1551–1588, November 1985. doi:10.1109/ISCAS.2003.1206020

49. Manolakis, D. G., Ingle, V. K., and Kogon, S. M. *Statistical and adaptive signal processing,* McGraw-Hill, New York, 2000.

50. Markel, J. D., and Gray, A. H., Jr. *Linear prediction of speech,* Springer-Verlag, New York, 1976.

51. Marple, S. L., Jr. *Digital spectral analysis,* Prentice-Hall, Englewood Cliffs, NJ, 1987. doi:10.1121/1.398548

52. Mitra, S. K. *Digital signal processing: a computer-based approach*, McGraw-Hill, New York, 2001.

53. Moon, T. K., and Stirling, W. C. *Mathematical methods and algorithms for signal processing*, Prentice-Hall, Upper Saddle River, NJ, 2000.

54. Oppenheim, A. V., and Schafer, R. W. *Discrete-time signal processing*, Prentice-Hall, Englewood Cliffs, NJ, 1999.

55. Papoulis, A. *Probability, random variables and stochastic processes*, McGraw-Hill, New York, 1965.

56. Papoulis, A. "Maximum entropy and spectrum estimation: a review," *IEEE Trans. Acoust. Speech Signal Process.*, vol. 29, pp. 1176–1186, December 1981.

57. Paulraj, A., Roy, R., and Kailath, T. "A subspace rotation approach to signal parameter estimation," *Proc. IEEE*, vol. 74, pp. 1044–1046, July 1986.

58. Peebles, Jr., P. Z. *Probability, random variables and random signal principles*, McGraw-Hill, New York, 1987.

59. Pisarenko, V. F. "The retrieval of harmonics from a covariance function," *Geophys. J. R. Astron. Soc.*, vol. 33, pp. 347–366, 1973. doi:10.1111/j.1365-246X.1973.tb03424.x

60. Proakis, J. G., and Manolakis, D. G. *Digital signal processing: principles, alogrithms and applications*, Prentice-Hall, Upper Saddle River, NJ, 1996.

61. Rabiner, L. R., and Schafer, R. W. *Digital processing of speech signals*, Prentice-Hall, Englewood Cliffs, NJ, 1978.

62. Rao, S. K., and Kailath, T. "Orthogonal digital lattice filters for VLSI implementation," *IEEE Trans. Circuits Syst.*, vol. CAS-31, pp. 933–945, November 1984.

63. Robinson, E. A. "A historical perspective of spectrum estimation," *Proc. IEEE*, vol. 70, pp. 885–907, September 1982.

64. Roy, R., and Kailath, T. "ESPRIT—estimation of signal parameters via rotational invariance techniques," *IEEE Trans. Acoust. Speech Signal Process.*, vol. 37, pp. 984–995, July 1989. doi:10.1109/29.32276

65. Rudin, W. *Real and complex analysis*, McGraw-Hill, New York, 1974.

66. Satorius, E. H., and Alexander S. T. "Channel equalization using adaptive lattice algorithms," *IEEE Trans. Commun.*, vol. 27, pp. 899–905, June 1979. doi:10.1109/TCOM.1979.1094477

67. Sayed, A. H. *Fundamentals of adaptive filtering*, IEEE Press, John Wiley & Sons, Hoboken, NJ, 2003.

68. Schmidt, R. O. "Multiple emitter location and signal parameter estimation," *IEEE Trans. Antennas Propagation*, vol. 34, pp. 276–280, March 1986 (reprinted from the Proceedings of the RADC Spectrum Estimation Workshop, 1979). doi:10.1109/TAP.1986.1143830

69. Schroeder, M. *Computer speech: recognition, compression and synthesis*, Springer-Verlag, Berlin, 1999.

70. Soong, F. K., and Juang B.-H, "Line spectrum pair (LSP) and speech data compression," *Proc. IEEE Int. Conf. Acoust. Speech Signal Process.*, pp. 1.10.1–1.10.4, March 1984.

71. Soong, F. K., and Juang B.-H, "Optimal quantization of LSP parameters," *IEEE Trans. Speech Audio Process.*, vol. 1, pp. 15–24, January 1993.

72. Strobach, P. *Linear prediction theory*, Springer-Verlag, Berlin, 1990.

73. Therrien, C. W. *Discrete random signals and statistical signal processing*, Prentice-Hall, Upper Saddle River, NJ, 1992.

74. Vaidyanathan, P. P. "The discrete-time bounded-real lemma in digital filtering," *IEEE. Trans. Circuits Syst.*, vol. CAS-32, pp. 918–924, September 1985. doi:10.1109/TCS.1985.1085815

75. Vaidyanathan, P. P. "Passive cascaded lattice structures for low sensitivity FIR filter design, with applications to filter banks," *IEEE Trans. Circuits Syst.*, vol. CAS-33, pp. 1045–1064, November 1986. doi:10.1109/TCS.1986.1085867

76. Vaidyanathan, P. P. "On predicting a band-limited signal based on past sample values," *Proc. IEEE*, vol. 75, pp. 1125–1127, August 1987.

77. Vaidyanathan, P. P. *Multirate systems and filter banks*, Prentice-Hall, Englewood Cliffs, NJ, 1993. doi:10.1109/78.382395

78. Vaidyanathan, P. P., and Mitra, S. K. "Low passband sensitivity digital filters: A generalized viewpoint and synthesis procedures," *Proc. IEEE*, vol. 72, pp. 404–423, April 1984.

79. Vaidyanathan, P. P., and Mitra, S. K. "A general family of multivariable digital lattice filters," *IEEE. Trans. Circuits Syst.*, vol. CAS-32, pp. 1234–1245, December 1985. doi:10.1109/TCS.1985.1085661

80. Vaidyanathan, P. P., Mitra, S. K., and Neuvo, Y. "A new approach to the realization of low sensitivity IIR digital filters," *IEEE Trans. Acoust. Speech Signal Process.*, vol. ASSP-34, pp. 350–361, April 1986. doi:10.1109/TASSP.1986.1164829

81. Vaidyanathan, P. P., Tuqan, J., and Kirac, A. "On the minimum phase property of prediction-error polynomials," *IEEE Signal Process. Lett.*, vol. 4, pp. 126–127, May 1997. doi:10.1109/97.575554

82. Van Trees, H. L. *Optimum array processing*, John Wiley & Sons, New York, 2002.

83. Wainstein, L. A., and Zubakov, V. D. *Extraction of signals from noise*, Prentice-Hall, London, 1962. doi:10.1119/1.1969250

84. Weiner, N. *Extrapolation, interpolation and smoothing of stationary time series*, John Wiley & Sons, New York, 1949.

85. Weiner, N., and Masani, P. "The prediction theory of multivariate stochastic processes," *Acta Math.*, part 1: vol. 98, pp. 111–150, 1957; part 2: vol. 99, pp. 93–137, 1958.

Author Biography

P. P. Vaidyanathan received his bachelor of science, bachelor of technology, and master of technology degrees from the University of Calcutta, India. He obtained his doctor of philosophy degree from the University of California at Santa Barbara and became a faculty of the Department of Electrical Engineering at the California Institute of Technology in 1983. He served as the executive officer of the Department of Electrical Engineering at Caltech for the period 2002–2005. He has received several awards for excellence in teaching at Caltech. He is the author of the book *Multirate Systems and Filter Banks* and has authored or coauthored more than 370 articles in the signal processing area. His research interests include digital filter banks, digital communications, image processing, genomic signal processing, and radar signal processing. He is an IEEE Fellow (1991), past distinguished lecturer of the IEEE SP society, and recipient of the IEEE ASSP Senior Paper Award and the S. K. Mitra Memorial Award (IETE, India). He was the Eliahu and Joyce Jury lecturer at the University of Miami for 2007. He received the F. E. Terman Award (ASEE) in 1995, the IEEE CAS Society's Golden Jubilee Medal in 1999, and the IEEE Signal Processing Society's Technical Achievement Award in 2001.

Index

Printed in the United States
by Baker & Taylor Publisher Services